The Science of Engineering Materials

The Science of Engineering Materials

by C. R. TOTTLE
M.MET., M.Sc., F.I.M., F.Inst.P., M.Inst.W.

Professor of Metallurgy, University of Manchester

CHEMICAL PUBLISHING CO., INC.
New York, N. Y.

The Science of Engineering Materials

ISBN: 978-0-8206-0096-3

Chemical Publishing Company:
www.chemical-publishing.com
www.chemicalpublishing.net

First American Edition:

© **Chemical Publishing Company, Inc.** - New York 1966

Second Impression:

Chemical Publishing Company, Inc. - 2011

Printed in the United States of America

Preface

This book is intended for engineers taking university and similar courses which include materials and as an introduction to materials science for metallurgists and those who require a general review. It is not intended to be a complete course in materials science nor to give details of commercial materials available.

In ancient times, the materials used for construction, and for weapons, tools, and ornaments, were limited to seven metals, stone, clay (dried and sometimes fired), earth, wood, animal hide and bone, and vegetable fibres woven into mats, twisted into ropes, or mixed with clay or earth. Little change in numbers or variety of materials occurred for thousands of years, but the technology of their use improved enormously. Bricks were made much stronger by better selection of clays, appropriate weathering and homogenizing, and higher temperature of firing. Alloys were developed from the traditional metals, although bronze had been known for some thousands of years. Nevertheless, wood continued to be the most flexible material for construction, possessing high strength-to-weight ratio, and capable of being joined by wood dowels, metal nails, or thongs of vegetable fibre or animal hide. The roof of the Tabernacle in Temple Square, Salt Lake City, Utah, is an excellent example of an unsupported large span formed from timber, joined only by dowels and thongs, and built just over one hundred years ago.

During the beginnings of scientific discovery, from the seventeenth century onwards, many materials, both elemental and compound, were explored, but very few indeed were of value in engineering. Charcoal continued to set the limit of very high temperatures attained on a large scale, until the Darby family carried out experiments to utilize a shaft furnace with coal previously purified by coking. Henry Cort's puddling process purified iron by utilizing coal fuel in a reverberatory furnace but could not melt the pure iron to separate the slag. An early appreciation of the value of ductility nevertheless stemmed from this association of low-carbon iron with slag in laminations, which allowed rolling into longer bars than had previously been possible and joining by hammer welding. The carrying capacity of structures increased as a result, and some composite structures were developed, to combine the best properties of more than one material. Many early locomotives employed sandwich frames of wood and iron.

One of the most significant steps in the development of steel occurred in 1740, when Benjamin Huntsman learned how to melt it in a coke-fired crucible furnace. The real advance, however, required an increase in the scale of operations. This became possible first by the Bessemer converter, and then by the Siemens–Martin open-hearth furnace. Steel remained a crude material until development of the deoxidation process for Bessemer steel by Mushet led to the addition of other alloying elements to modify properties. Men like Whitworth were then able to develop machine tools and, in turn, precision engineering, employing more stable materials exhibiting a wide variety of properties as a result of heat treatment. Non-ferrous metals did not advance so rapidly until the commercial exploitation of aluminium, which subsequently contributed so much to the aeroplane. In recent times, almost every metallic element in the periodic table has been prepared on a commercial scale and has been investigated as a possible engineering material.

Chemical compounds in profusion followed the industrial revolution and its new demands. Portland cement made possible bricks of larger size and different shapes and monolithic structures in complex forms. In more recent times, new fibres have replaced many traditional ones, not only for clothing but also for reinforcement of plastics, replacement of metal wires, etc. The plastics industry alone offers an immense variety of materials, most of which can be 'tailor-made' for the conditions of service. The electrical industry replaced porcelain and similar ceramic insulators for indoor use with the phenol formaldehyde resins at an early stage. Before the second world war, not only these resins, but also the urea formaldehyde types, the transparent acrylics and the cellulose esters, were commercially available and applied in engineering. In the twenty-five years following, an enormous quantity of synthetic polymer materials has replaced much of the metal in the domestic kitchen, leather, wool, and cotton in furnishings, and natural rubber, wood, and metal in vehicles, prime movers. Plastics are still limited for high-temperature applications, but the so-called ceramic materials, normally metal oxides, carbides, silicides, nitrides, and borides, are likely to supersede metals to a considerable extent in this field.

Electrical developments have given rise to the utilization of hitherto uninteresting elements such as silicon and germanium, and superconducting alloys are likely to provide an outlet for some intermetallic compounds of little mechanical significance. Faced with such sweeping developments and with changes in demand, the design engineer can no longer limit himself to traditional materials, nor to the traditional reliance on catalogues of properties or manufacturers' specifications. Interactions between widely diverse properties may be involved in new developments, and more stringent service conditions are continually to be expected. The lifetime of a material is not always necessarily long. A manned space craft may have a lifetime of a few hours, the rocket that carries it aloft even less. Power-station generators may be expected to last for twenty to twenty-five years, motor-cars for only ten. Economic considerations obviously enter into this argument, but knowledge of the properties and the effect of time are far more restricting than they were, and generous safety factors are becoming less welcome.

To cope with this situation, an understanding of the basic science that explains the variation in properties between elements, compounds, alloys, and mixtures becomes essential. The modification of the fundamental properties by manufacturing operations and subsequent service conditions has to be superimposed on this background, but the detailed consideration is really the realm of the metallurgist, chemist, or physicist. The engineer can, in these days, normally draw on the services of these materials scientists for this detail, and so requires a less detailed approach to individual materials and more of the broad overall picture of those which are feasible.

This book attempts to provide the broad background, to illustrate the basic reasons for the properties of elements, and to explain the consequences of chemical combination, alloying, and mixing. Most previous books have touched only lightly on the atom itself, but my experience in teaching engineers in the University of Manchester suggests that a greater depth of approach is welcome, perhaps because it can account for so much of the subsequent behaviour of metals. I acknowledge the advice and comment of my engineering colleagues in this matter, particularly Professors J. Diamond, M. R. Horne, and W. B. Hall, and their academic staff. Inevitably, such an approach has its own limitations. This book does not deal in any *detail* with metallic alloy systems and their equilibrium, nor with the specialized metal compounds used as engineering materials. It does, however, compare and contrast them in terms of structure and the dependence of properties on the atomic configuration. Books produced especially to deal with the details of specific classes of materials are best consulted in the light of this scientific background.

It is also my hope that many engineers of long standing may find in this book some explanation of the apparent perversity of certain materials, some confirmation of the reluctance of materials scientists to commit themselves in certain respects, and a more tolerant attitude to producers and research departments alike. I have in mind particularly the chief design engineer who refused to believe that temperatures in the cylinders of a compression-ignition oil engine reached the level indicated by microscopic examination of a fissured piston crown. Twenty years ago he wrote 'rubbish' across my report to this effect, but I did have the satisfaction some months later of noting that my recommendation for replacement of the material by a low-alloy steel had been implemented in the machine shops. Today there are alternative solutions to this and other problems, and no doubt design engineers are more willing to examine each possibility. May this book help them to clarify the whole picture.

Manchester, April 1964 C. R. Tottle

Acknowledgments

I would like to express my sincere thanks to Dr Michael Lucas of the Argonne National Laboratory, Illinois, for his considerable help in reading the proofs. I also acknowledge my debt to the authors of the books recommended for further reading, on whose ideas I have drawn in the past when preparing the lectures on which this book is based.

Western Springs, Illinois C. R. T.

May 1965

Contents

List of Plates

1

The Free Atom

1.1. *The contribution of spectroscopy*

The original work on the nature of substances was carried out by chemists, who were anxious to explain the variation in behaviour of the various elements and their compounds. John Dalton, for example, established the existence of atoms, and defined them as the smallest part of an element that could exist alone and still be representative of that substance. Later work by physicists delved into the atom itself, and demonstrated the existence of elementary particles (chiefly electrons and protons), which varied in number with the element concerned. At this stage, however, the internal arrangement of these particles was inadequately understood, and it was not possible to correlate the properties of atoms with their structure. The investigation of the radiation emitted by excited atoms and the diffraction of X-radiation by atoms gradually changed this situation. Kirchhoff first suggested in 1859 that each species of atom has a uniquely characteristic spectrum. He and Bunsen analysed the sun's atmosphere in 1861 and discovered two alkali metals (rubidium and caesium) by investigating atomic spectra.

The spectrometer used in such techniques allowed the light from the source (high-temperature flame, gas-discharge tube, electric arc, or high-frequency spark) to pass through a narrow slit into a collimator, and thence through a prism, where refraction and dispersion took place, and finally via a telescope to an eyepiece for visual examination, or to a photographic plate. A series of lines resulted, one for each wavelength of radiation emitted, each line being an image of the slit produced by that particular wavelength. The more complex the atom, the more wavelengths were emitted, and thus the greater the number of lines produced.

Utilizing an electric discharge through a glass tube filled with a pure gas, the spectra of gaseous elements can also be investigated. In 1885 BALMER discovered a relationship between the lines produced as part of the hydrogen spectrum. He expressed the wavelength of the lines as

$$\lambda = \frac{3{,}646{\cdot}13m^2}{m^2 - 4}\ \text{Å} \tag{1.1}$$

where m has values 3, 4, 5, 6 . . . ∞.

These measurements were made in air, and thus they are not perfectly reproducible, but by changing to vacuum measurement and using $\lambda_{vac} = 3{,}647{\cdot}16m^2/(m^2 - 4)$ they can be reproduced anywhere in the world. Thus, when $m = 3$, $\lambda = 6{,}563$ Å; when $m = \infty$, $\lambda = 3{,}647{\cdot}16$, giving the longest and shortest wavelength lines in the series, known as the *Balmer series.* In spectroscopy, it is more customary to use the wave number

$$\nu = \frac{n}{c} = \frac{\text{Frequency}}{\text{Velocity}} = \frac{1}{\text{Wavelength}}$$

i.e. the number of waves per centimetre. Rewriting the Balmer series in this way, and rearranging the equation,

$$\nu = \frac{109{,}675}{2^2} - \frac{109{,}675}{m^2} \tag{1.2}$$

where $m = 3, 4, \ldots \infty$. RYDBERG, working on the spectra of alkali metals, followed Balmer's work in predicting further series of lines covered by a simple formula. Rydberg and RITZ also predicted that the hydrogen spectrum should contain other series, given by a formula similar to that of the Balmer series. Subsequent experiment established these series, namely:

$$\nu = \frac{109{,}675}{1^2} - \frac{109{,}675}{m^2}$$

where $m = 2, 3, \ldots \infty$ (discovered by LYMAN in the ultraviolet).

$$\nu = \frac{109{,}675}{3^2} - \frac{109{,}675}{m^2}$$

where $m = 4, 5, 6, \ldots \infty$ (discovered by PASCHEN in the infrared).

At a later stage, improvements in instruments led to yet one more series (due to BRACKETT) which occurs in the far infrared of the hydrogen spectrum:

$$\nu = \frac{109{,}675}{4^2} - \frac{109{,}675}{m^2}$$

where $m = 5, 6, 7 \ldots \infty$.

These four series embrace the normal line spectrum of hydrogen, from infrared (long wavelength) to ultraviolet (short wavelength). Rearranging these equations covering the hydrogen spectrum in order

$$\nu = \frac{R}{(1+0)^2} - \frac{R}{(m+1)^2} \quad \text{(Lyman)}$$

$$\nu = \frac{R}{(1+1)^2} - \frac{R}{(m+2)^2} \quad \text{(Balmer)}$$

$$\nu = \frac{R}{(1+2)^2} - \frac{R}{(m+3)^2} \quad \text{(Paschen)}$$

$$\nu = \frac{R}{(1+3)^2} - \frac{R}{(m+3)^2} \quad \text{(Brackett)}$$

where R is 109,675 cm^{-1}, the Rydberg constant, and m is now an integer varying from 1 to ∞.

It is obvious that we can write a general equation covering these four series in which there is a limiting term and a variable term, the limiting term being constant for any one series but increasing in steps as we move from short wavelength (Lyman) to long wavelength (Brackett). This 'combination' principle was first proposed by Ritz. The limiting term in the Lyman series for hydrogen is $R/1^2 = R$, and the first variable term will be $R/(m + 1)^2 = R/2^2$. The Balmer series has a limiting term $R/2^2$, which is the first variable term of the shorter wavelength Lyman series. The first variable term in the Balmer series is $R/3^2$, which in turn becomes the limiting term of the next, the Paschen, series. Likewise the first variable term of the Paschen series is $R/4^2$, and this is the limiting term of the Brackett series. There is obviously a continuous connection, through these series, of the factors affecting the emission of spectral lines. The denominator of the limiting terms is given by squaring the sum of a constant (1) plus an increasing integer (0, 1, 2, or 3). The denominator of the variable term is similarly made up by squaring the sum of a variable integer ($m = 1, 2, 3 \ldots \infty$) plus an increasing integer (1, 2, 3, or 4). Rewriting this as a general equation

$$\nu = \frac{R}{(m'+a')^2} - \frac{R}{(m+a)^2} \tag{1.3}$$

For hydrogen

$m' = 1$,	for each series
a' has values 0, 1, 2, 3,	through the four series
$m = 1, 2, 3 \ldots \infty$,	for each series
a has values 1, 2, 3, 4,	through the four series

and substitution in the above equation would give all the series of spectral lines for hydrogen. The values of a' and a will not of course be the same when the general equation is applied to the spectra of all other elements, as they are in hydrogen, but the appropriate substitution will give the wave numbers for all spectral lines. Rydberg also observed that the lines in spectra were characterized by certain other properties, quite independent of the atom concerned. Certain of the lines were always very sharply defined: these he called 'sharp'; certain were 'diffuse'. Some lines appeared even when the element concerned was present in only small quantity, and these were given the name 'principal'. A fourth type was recognized later, and called 'fundamental'. These features, Rydberg suggested, were accounted for by the same close relationship between the series of spectral lines that had been observed for hydrogen, and therefore one could write this into the general equation by substitution for one of the variable integers. He suggested that a in the general equation (1.3) took the values s, p, d, f, where these letters stood for sharp, principal, diffuse, and fundamental. The four variable terms would then be

$$\frac{R}{(m+s)^2};\quad \frac{R}{(m+p)^2};\quad \frac{R}{(m+d)^2};\quad \frac{R}{(m+f)^2}$$

In order to account for the relationship between the first variable term of the lowest series (Lyman for hydrogen) and the limiting term of the next (Balmer for hydrogen) Rydberg proposed a so-called 'selection rule'. Writing S–P–D–F to stand for the four series, he proposed that any one series, written in this order, was related to its immediate neighbours, but not with distant ones. Thus S can be related only to P; P to S or D; D to P or F; F only with D. In writing his general equation therefore, the selection rule implied that the limiting term of the sharp series was the first variable term of the principal. Hence,

$$\nu = \frac{R}{(2+p)^2} - \frac{R}{(m+s)^2} \tag{1.4}$$

where $m = 1 \ldots \infty$. This gives the SHARP series of lines.

Accordingly, the first variable term of this sharp series becomes the limiting term of the PRINCIPAL series

$$\nu = \frac{R}{(1+s)^2} - \frac{R}{(m+p)^2} \tag{1.5}$$

where $m = 2 \ldots \infty$.

The other two series follow suit:

DIFFUSE:

$$\nu = \frac{R}{(2+p)^2} - \frac{R}{(m+d)^2} \tag{1.6}$$

where $m = 3 \ldots \infty$.

FUNDAMENTAL:

$$\nu = \frac{R}{(3+d)^2} - \frac{R}{(m+f)^2} \tag{1.7}$$

where $m = 4 \ldots \infty$.

The lowest terms of the sequences are therefore

$$\frac{R}{(1+s)^2}; \quad \frac{R}{(2+p)^2}; \quad \frac{R}{(3+d)^2}; \quad \frac{R}{(4+f)^2}$$

To avoid writing this term in full, it is simpler to use a shorthand version, where mS stands for $R/(m+s)^2$. Hence 1S means $R/(1+s)^2$ the lowest term of the sharp series. The other lowest terms are then 2P; 3D; 4F.

Provided that values of s, p, d, f are available for measurements of spectral lines, all other lines that exist should be predictable by substitution in these general equations for the various series. This proves to be so.

NIELS BOHR was the first to recognize a possible relationship between these series of spectral lines and the structure of the atom. He selected the hydrogen atom as his model, partly because of its simplicity (one proton in the nucleus and one extra-nuclear electron) and partly because the spectral series were best known for hydrogen.

1.2. *Bohr's early theory*

Bohr envisaged the electron in a hydrogen atom revolving round the nucleus in a planetary motion, in a fixed circular orbit. If external energy were applied to the atom, there would be an increase in the orbit radius. When, subsequently, the electron gave up this excess energy and returned to its normal orbital motion, the excess energy would appear as radiation. Thus, excitation of a hydrogen atom might increase the energy level of the electron temporarily, with a subsequent return to the 'ground state' of energy with the emission of radiation—detected by a spectrometer as a spectral line. If the energy level of the ground state of the electron is W_0, and that of the excited state W_e, then the difference must refer to the energy of the radiation given out, namely,

$$\Delta W = W_e - W_0 \tag{1.8}$$

where ΔW = work done in emitting radiation. PLANCK had meanwhile argued that vibrational energy was never absorbed or emitted in a continuous manner, but in discrete bundles, which he called 'quanta'. He postulated a *constant of action*, equal to the product of energy and time. Where spectral lines are involved, the emitted frequency is concerned also; hence

$$\text{Action} \times \text{Frequency} = \frac{\text{Action}}{\text{Time}} = \frac{\text{Energy} \times \text{Time}}{\text{Time}} = \text{Energy}$$

Thus the value of Planck's constant of action h multiplied by frequency n gives us a basic measure of energy hn. This is the basic unit of energy, the quantum.

This reasoning can be applied to the energy involved in emitting spectral lines. Substituting in the general Rydberg equation (1.3),

$$\nu = \frac{R}{m_1^2} - \frac{R}{m_2^2}$$

where m_1 and m_2 are integers as previously defined, $\nu = n/c$. Hence

$$n = \frac{cR}{m_1^2} - \frac{cR}{m_2^2} \quad \text{and} \quad hn = \frac{hcR}{m_1^2} - \frac{hcR}{m_2^2} \tag{1.9}$$

This gives an expression for one quantum of energy, in terms of the three constants h, c, R, and the inverse square of two integers, one quantum fixed, the other variable, for any given series of spectral lines.

There is a dimensional resemblance between the two equations (1.8) and (1.9)

$$hn = \frac{hcR}{m_1^2} - \frac{hcR}{m_2^2} \quad \text{and} \quad \Delta W = W_e - W_0$$

If ΔW is related to quanta, Bohr envisaged a situation in which one quantum of energy is involved, and $\Delta W = hn$. Then, in general terms,

$$W = \frac{hcR}{\bar{m}^2} \tag{1.10}$$

where W is the energy level of an electron orbit and $\bar{m}$ is an integer. This would mean that energy levels of electron orbits were related inversely to the squares of natural numbers.

Bohr set out to calculate energy levels in the hydrogen atom, in terms of the forces involved, and showed that

$$W = -\frac{eZe}{2a}$$

where e is the charge on the electron, Ze the charge on the nucleus, and a the 'radius' of orbit.

Thus $a = eZe/2W$. Substituting from equation (1.10),

$$a = \frac{eZe\bar{m}^2}{2hcR} \tag{1.11}$$

Thus, the radii of orbits would be directly related to the squares of natural numbers. In moving from one orbit to another, the energy involved is measured in units involving hn, i.e. quanta. The minimum energy change $1 \times hn$ (or $h\nu c$) can thus be multiplied to apply to many different energy levels of electrons in orbits, or the energy given out when the electron returns to a lower energy state. The line spectra are a direct result of these energy changes, and the equations for the series of spectral lines are merely the expression of the minimum and maximum energy levels of the electrons in the atom concerned, with the individual steps between these in units of hn. Once an electron goes beyond the maximum energy level it is lost to infinity, the atom is said to be 'ionized'. A whole new series of spectral lines may then result from the ionized atom, because the balance of charges, and therefore energies, is upset. Later the ionized atom will recapture an electron and return to neutrality.

Spectroscopy then outlined the principles of energy levels and led to a crude model of the simpler atoms. The complex atoms were more difficult, but the general principles still applied.

These ideas were extended to envisage the electron revolving around the nucleus with angular velocity ω, so that on moving into the next inner orbit, and emitting a spectral line as a result, the energy change could be equated to the change in angular momentum of that electron, in addition to the frequency of radiation emitted. There was no evidence for this assumption, but it formed a basis for further elucidation of the motion of electrons. Bohr proceeded to define a 'quantum number' k referring to this angular momentum of the electron, to which he gave values from 0 to 3, corresponding to the values s, p, d, f in the general Rydberg equations governing the series equations (1.4)–(1.7). The Rydberg selection rule thus refers to a transition between energy levels when $\Delta k = \pm 1$.

We now have two quantum numbers by which to define the orbit of the electron and energy state. The principal quantum number n defines the general energy level, and is the natural number m that we have used in the equations governing spectral lines. The radii of the orbits are proportional to n^2 (1.11) and the energy levels to $1/n^2$ (1.10). The secondary quantum number k is the measure of angular momentum mentioned above, this angular momentum being $kh/2\pi$. In circular orbits n and k are equal, but in elliptical ones the ratio of k/n gives the ratio of the minor to the major axis of the ellipse.

k then has values 1, 2 . . . n. At this stage, Bohr's theory ceased to be sufficiently explicit. It defined the electron motion in the terms of classical dynamics, yet accepted the quantum restrictions that apply to wave motion. Just as radiation is now regarded as neither purely corpuscular (obeying classical mechanics) nor as purely a wave motion (obeying wave mechanics) so the electron cannot be envisaged as rotating in orbits nor as simply an electromagnetic vibration. The modern approach is to define the probability of an electron being in a certain location or occupying a given energy level. The adaptation of classical mechanics to wave theory dates back only some forty years, so that modern concepts of the structure of atoms are very new, and yet can be quite effectively applied.

De Broglie and Schrödinger first showed that a particle of mass m and velocity v could be associated with a wave, given by $\lambda = h/mv$. This was confirmed by the discovery of electron diffraction, whereby a stream of electrons behaves similarly to visible light. (It is also the basis of electron microscopy.) The atom is then regarded as a nucleus surrounded by a cloud of negative electricity, the density of which is proportional to the probability of finding an electron at a particular point. This probability is defined by a wave function, the square of whose amplitude is a measure of that probability. The normal energy states of the electrons are still related by the squares of natural numbers, as in the Bohr theory, and similar quantum numbers are retained to secure definition of these states.

1.3. *Modern quantum theory of the atom*

Four quantum numbers are required, defined as follows:

The principal quantum number n is numerically equal to that defined by Bohr, and is a measure of the electron energy in elementary terms.

The secondary quantum number l is a measure of the angular momentum of the electron. It is numerically equal to $k - 1$ of the Bohr theory, and l may have any value from 0 to $n - 1$. A value of $l = 0$ does not mean an electron at rest, but a motion that does not give rise to a resultant angular momentum. As with the Bohr theory, the possible values of l (0, 1, 2, 3) are given the identification s, p, d, f. Taken with the values of n, the various energy levels expressed in terms of the principal and secondary quantum numbers correspond to the electron states used in spectroscopy, e.g. $n = 1$, $l = 0$ means $1s$, and $n = 3$, $l = 2$ means $3d$. The angular momentum of the electron is of magnitude $(h/2\pi)\sqrt{[l(l+1)]}$, whereas Bohr's theory gave simple multiples of $h/2\pi$.

In dealing with atoms, we are concerned with moving electric charges, which inevitably generate magnetic fields. The electrons in motion have associated with them a magnetic moment, owing to the orbital angular momentum. This magnetic moment is given by the expression $(eh/4\pi mc)\sqrt{[l(l+1)]}$ for orbital motion, where l is the charge on the electron and m its mass. The quantity $eh/4\pi mc$ is called the *Bohr magneton*, a convenient unit of electron magnetic moment.

If the electron is subjected to an applied magnetic field, then the angular momentum, which as we have seen gives rise to a magnetic field itself, must be affected by the external field. The result is defined by using a third quantum number m_l taken to be a measure of the component of angular momentum in the direction of a superimposed weak magnetic field. The component in this direction is then $\frac{1}{2}hm_l$, and, similarly, the component of magnetic moment in the direction of an applied field is then $ehm_l/4\pi mc$. The quantum number m_l may have any value from $+l$ to $-l$, including 0.

There is one final feature, which produces changes in spectral lines when the atoms are excited in the presence of an applied magnetic field. The effects can be explained if it is assumed that each electron is spinning on its own axis, and that the spin produces a magnetic moment. The magnitude of this magnetic moment is given by

$$\frac{2e}{2mc}\frac{h}{2\pi}\sqrt{[s(s+1)]}$$

where $s = \frac{1}{2}$. The component of this magnetic moment in the direction of the field is $ehm_s/2\pi mc$. The symbol m_s is the fourth quantum number, the *spin quantum number*, and is equal to $\pm\frac{1}{2}$, to fit the equation of magnetic moment above. The $\pm$ sign is necessary to indicate the spin direction when more than one electron is involved, i.e. whether the electrons are parallel or opposite in their spin direction.

Thus, the modern quantum theory requires definition of all four quantum numbers n, l, m_l, and m_s, to fix any energy level in which there is a probability of finding an electron. It must be remembered that even in a simple atom such as hydrogen, with only one electron, it is necessary to study not only the ground state of energy of that electron, i.e. the level it occupies when in the state of lowest energy, but also the levels into which it may be excited by increasing its energy, and from which it may return accompanied by the emission of spectral lines. In the hydrogen atom, this can be illustrated by taking the lowest energy state first, and then examining the meaning of the four quantum numbers as electron energy is increased.

The lowest possible value of n is 1, and that of l is 0. This is known as the 1s state of the electron, since $n = 1$ and $l = 0 = s$. There is no need at this stage to consider m_l, the component in a magnetic field (for since $l = 0$, $m_l = 0$), nor the electron spin, since only one electron is involved. It can be shown that the s states of all electrons, i.e. those in which the value of l is 0, are spherically symmetrical, with $n - 1$ spherical nodes, i.e. regions where there is no chance of finding an electron. Thus, the ground state of hydrogen, where $n = 1$, has no spherical nodes, i.e. there is no region within the basic sphere

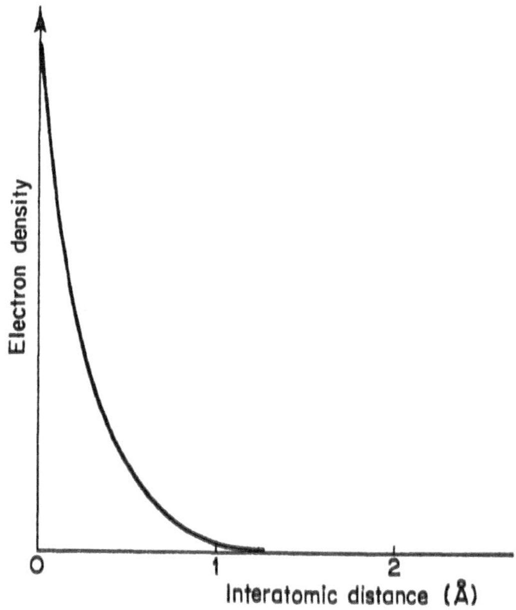

FIG. 1.1. The electron density of the (1s) state in hydrogen (after Hume-Rothery).

of symmetry where one is unlikely to find the electron. The electron of hydrogen may therefore be regarded as contained in a spherical ball of negative electricity which is of high density close to the nucleus. If a hydrogen atom could be seen and the electron made to 'glow', then the general appearance would be of a diffuse sphere of light whose density varied as in Figure 1.1 and Plate Ia. Thus, the electron really has a three-dimensional motion, and the Bohr orbit has here become a sphere of probability.

If we now increase the energy of this electron, say that of the 2s state $n = 2$, and $l = 0$, then the electron still occupies the spherical symmetry associated with s states, but now with one spherical node ($n - 1 = 2 - 1$). This is illustrated in Plate Ib. Similarly when $n = 3$ and $l = 0$, the 3*s* state, there will be $n - 1 = 2$ spherical nodes, as shown on Plate Ic. Increase in energy also arises by moving to the *p* states of the hydrogen atom, i.e. where $l = 1$. The lowest value of *n* here is 2, the 2*p* state. Since $l = 1$, there is now a possible variation in m_l, since $m_l = +l$ through 0 to $-l$. In fact m_l has three values, $+1$, 0, and -1. The *p* states have a node at the centre, i.e. there is zero probability of an electron being there, and they have $n - 2$ spherical nodes. The value of m_l is here significant, because it determines the direction of the electron density in fulfilling the requirements of the *p* states. Since there are $n - 2$ spherical nodes, then for $n = 2$ (2*p* state) there will be no spherical nodes, and as with the 1*s* state for hydrogen, the density will be vanishingly small at some distance from the nucleus. To indicate the values of m_l and the effect of applied magnetic field, a model can be constructed (Plate II) showing the possible density of the electron cloud for the three values of m_l. It can be seen that the electron may occupy any one of these three 'egg-like' cloud shapes, according to the m_l value, i.e. the direction of an applied magnetic field. All of these are 2*p* states, but the magnetic field in one of three mutually perpendicular directions will affect the exact orientation of the shape occupied by the electron.

In the *d* states, i.e. when $l = 2$, the lowest state is when $n = 3$, the 3*d*. Here m_l can be $+2$, $+1$, 0, -1, or -2, and there are five directional effects to be considered. There will be a node at the centre and $n - 3$ spherical nodes. For the *f* states, when $l = 3$, the lowest state is $n = 4$, the 4*f*, and there are seven possible directional effects since m_l has seven values, with $n - 4$ spherical nodes. The representation of such complex electron states by models or drawings is almost impossible, but the general inference of symmetry is not too dissimilar from that for the *p* states.

With more complex atoms than hydrogen there are more electrons, and there may be interaction between these electrons, especially when some of the outer ones are shielded from the nucleus by those of lower energy.

1.4. *The exclusion principle*

Just as selection rules were applied under the older theory of Bohr to account for the relationship between series of spectral lines, so it is necessary to define the possible occupancy of energy states in complex atoms under the newer quantum theory. Because of the interaction referred to, the lowest energy state for several electrons cannot be achieved by assigning to them the lowest values of the quantum numbers n and l. The PAULI principle states that not more than one electron can be in a given energy state as defined by all four quantum numbers. It is possible of course to vary m_l (once $l = 1$) and subsequently m_s, for any set of values of n, l, and m_l, because two electrons, of opposite spin, can occupy the same n, l, m_l combination. The maximum number of electrons allowed by the Pauli exclusion principle in terms of these quantum numbers is therefore:

(1) As the principal quantum number n varies 1, 2, 3, 4 . . ., then the number of electrons that can be contained is 2, 8, 18, 32, respectively, or $2n^2$.
(2) For a given value of n, the maximum number of electrons when $l = 0$, the s state, is 2; for $l = 1$, the p state, 6; for $l = 2$, the d state, 10; and for $l = 3$, the f state, 14 electrons.

It is now possible to build up the periodic table of elements in terms of the atomic number, i.e. the number of extra-nuclear electrons, by evaluating the four quantum numbers. By this means, the similarity of properties revealed in the periodic system is reflected in terms of the electron energy levels and effects such as the electron spin and the magnetic moment due to the angular momentum of the electrons. In all these considerations, the 'free' atom is being considered, i.e. the atom as a single unit, as in spectroscopy or in chemical solution, and not the atoms bound together in the solid state. The solid state is derived from the free atom configuration by taking into account the different ways in which binding can take place. This aspect is discussed at length in Chapter 2.

1.5. *The periodic system*

As noted many times, above, hydrogen, with an atomic number Z of 1, has one electron, in the $1s$ state, the lowest possible energy state, and nearest to the nucleus. This is written as $(1s)^1$ to indicate 1 electron in the $1s$ state.

The next atom, helium He, has two electrons ($Z = 2$), and the Pauli principle states that only one electron can exist in which $n = 1$, $l = 0$ ($m_l = 0$) unless we vary the electron spin. Since $m_s = \pm\frac{1}{2}$, there are two opposite spin directions, the two electrons are in the $1s$ state $(1s)^2$ but of opposite spin. These two electrons completely fill the $1s$ state, so that in the next atom, lithium Li ($Z = 3$), the first two electrons occupy the $1s$ state as in helium, but the third must change in quantum state. The obvious change is $n = 2$, retaining $l = 0$, i.e. the $2s$ state. Thus the electronic configuration of lithium is written $(1s)^2(2s)^1$.

Beryllium Be ($Z = 4$) can have two electrons in the $2s$ state, since they can have opposite spin; hence the electron configuration is $(1s)^2(2s)^2$ and there are no further possibilities for the $2s$ state.

Accordingly with boron ($Z = 5$) the fifth electron must enter a new subgroup and can do so by adopting the state $l = 1$, i.e. the $2p$ state. Boron is thus $(1s)^2(2s)^2(2p)^1$. From this point, the $2p$ state can build up to its maximum of six electrons, since $m_l = +1, 0, -1$, and there are the two values of electron spin for each value of m_l.

Carbon ($Z = 6$)	is thus $(1s)^2(2s)^2(2p)^2$	$m_l = +1, -1$
Nitrogen ($Z = 7$)	is thus $(1s)^2(2s)^2(2p)^3$	$m_l = +1, 0, -1$
Oxygen ($Z = 8$)	is thus $(1s)^2(2s)^2(2p)^4$	$m_l = +1$ with $m_s = \pm\frac{1}{2}$; $m_l = 0$; $m_l = -1$
Fluorine ($Z = 9$)	is thus $(1s)^2(2s)^2(2p)^5$	$m_l = \pm 1$ with $m_s \pm\frac{1}{2}$; $m_l = 0$
Neon ($Z = 10$)	is thus $(1s)^2(2s)^2(2p)^6$	$m_l = \pm 1$ and 0 and $m_s \pm\frac{1}{2}$.

It is now necessary to make a decision on whether the first variant, from boron to nitrogen, is made by varying only m_l, followed later by the additional electron spin change, or whether the spin changes for each value of m_l once the latter has changed. From the magnetic properties it can be shown that normally the electron spin tends to remain parallel as long as possible as atomic number increases, so that m_l varies first, and m_s only when the possibilities are exhausted. This shows up in the magnetic properties of oxygen, which is the first element with a pair of electrons having antiparallel spin. At neon the stable configuration of all states involved is reached, as with helium. The inert properties and lack of physical peculiarities such as magnetic influences in the rare gases can thus be accounted for by the completion of electron groups as defined by the four quantum numbers.

From neon to sodium ($Z = 11$) a new group must start, with $n = 3$, $l = 0$, or $(3s)^1$. Sodium is thus like lithium, which had $(2s)^1$ and hydrogen with $(1s)^1$. (Hydrogen is exceptional, however, as the simplest element.) Lithium and sodium are alkali metals of high reactivity and high conductivity, and the significance of this will become more evident in Chapter 2.

From sodium the type of variation seen in the first period is repeated, so that magnesium has $(3s)^2$ for its outer electrons. Aluminium ($Z = 13$) must again form a p subgroup, and has $(3s)^2(3p)^1$.

Silicon $(3p)^2$, phosphorus $(3p)^3$, sulphur $(3p)^4$, chlorine $(3p)^5$, and argon $(3p)^6$ follow suit, and along the same lines as their counterparts in the first period. Again the tendency for parallel spin applies, so that m_l goes through its variations before m_s needs to change. Sulphur, where antiparallel spin enters the $3p$ subgroup, should behave like oxygen, which showed the same phenomena in the $2p$ group. This is so, and the particular magnetic properties of sulphur can be accounted for by this change in electron spin. One aspect of importance is the occurrence of many of the true non-metals in these two short periods, all of which have electrons occupying the $2p$ or $3p$ subgroups. (As demonstrated later, the elements that have $4p$ and $5p$ electrons are also either non-metallic or have peculiar properties not entirely metallic in nature.) In the third period, potassium ($Z = 19$) and calcium ($Z = 20$) continue the build-up observed before with sodium and magnesium, i.e. commencing another s group, this time with $n = 4$—hence the $4s$ electrons, potassium having one and calcium two. These are respectively monovalent and divalent elements, calcium having its two $4s$ electrons with opposite spin. If the normal development of the $4p$ electrons were to follow, as with the first two short periods of the system, then scandium ($Z = 21$) would follow calcium and have its outer electrons distributed as $(4s)^2(4p)^1$. However, the energy levels above $4s$ are of closely similar magnitude, and it so happens that $4s$ is only slightly lower in energy level than $3d$, which in turn is just less than $4p$. Scandium adopts the $3d$ level, i.e. $n = 3$, $l = 2$, before the $4p$. Having inserted this fresh group of electron energy levels, the $3d$, we can have a maximum of ten electrons while continuing to satisfy the Pauli principle, because $l = 2$, so that m_l has five values, $+2$, $+1$, 0, -1, -2, and each value of m_l has the two possible values of m_s as usual. From scandium to nickel ($Z = 28$) the $3d$ levels should fill up to eight electrons in a straightforward manner, Sc $(3d)^1$, Ti $(3d)^2$, V $(3d)^3$, Cr $(3d)^4$, Mn $(3d)^5$, Fe $(3d)^6$, Co $(3d)^7$, Ni $(3d)^8$. Since once more there is a variation in m_l with opposite or parallel spin electrons, it is necessary to specify whether m_l varies first, followed by the spin variation, or *vice versa*. Again, it appears that m_l varies and the spin tends to remain parallel, so that m_s does not need to vary until we reach iron ($Z = 26$). Iron has one pair of opposite spin and four electrons with parallel spin. Cobalt shows two pairs with opposite spin and three electrons with parallel spin. Nickel has three pairs of opposite spin and two parallel. In these elements, the gradual change in electron spin characteristics causes a change in magnetic properties, exactly as it did in those elements of the earlier periods with build-up of the p electrons. In this case, however, the fact that there are five values of m_l possible, with parallel spin, gives rise to a more pronounced effect, which will be discussed in greater detail in Chapter 6. A further difference in properties arises from the closeness of electron energy levels between $4s$, $3d$, and $4p$. This proximity of energy allows more ready transfer of one or more electrons between these levels, and so accounts for the variable valency displayed in chemical reactions with these elements. As a result of these quite marked changes in magnetic and chemical properties in this, the first long period of the table, these elements are known as the *transition elements*, because they fill in the space between the $4s$ and $4p$ electron types. The two elements following nickel, copper ($Z = 29$), and zinc ($Z = 30$), which should complete the $3d$ electron state to its maximum of ten electrons, show the expected trends in properties, but not to such a marked degree, since completion of the $3d$ level balances out the electron spins into pairs. Thus from scandium to zinc a gradual build-up to the maximum parallel spin at manganese $(3d)^5$ is followed by a gradual build-up to opposing spin balance from iron $(3d)^6$ to zinc $(3d)^{10}$. Copper has two valencies, and the normal electron configuration is now believed to be $(4s)^1(3d)^{10}$ for the monovalent state rather than $(4s)^2(3d)^9$. Similarly, the chemical properties of chromium indicate a configuration of $(4s)^1(3d)^5$ rather than $(4s)^2(3d)^4$. Such variations again illustrate the close proximity of energy levels in the transition elements.

After zinc, we move to gallium ($Z = 31$), where the $4p$ state is now the next lowest energy level, and we revert back to the previous behaviour of elements in the short periods. The $4p$ group builds up to its maximum of six electrons through germanium Ge ($Z = 32$) to krypton Kr ($Z = 36$). Each of these elements has the $(4s)^2(3d)^{10}$ inner core of electrons built up in the transition elements. This completes the first long period, with bromine falling under chlorine as the third halogen, and krypton under argon as another inert gas. The second long period now commences with a repetition of the first, i.e. two $5s$ electrons, ten $4d$, and six $5p$.

Rubidium ($Z = 37$) is like sodium, with $(5s)^1$, and the following element, strontium ($Z = 38$), resembles calcium, having $(5s)^2$. Once more a set of transition elements follows, filling $4d$ from yttrium $(4d)^1$ to cadmium $(4d)^{10}$. These are followed by elements filling the $5p$ level, from indium $(5p)^1$ to xenon Xe $(5p)^6$. Table I shows the detail of this and previous periods, to which the argument used above can be applied in completion of the whole periodic system.

The third long period commences in a similar fashion, with caesium ($Z = 55$) having $(6s)^1$ and barium ($Z = 56$) $(6s)^2$. Lanthanum follows, ($Z = 57$) with the first $5d$ electron, just as for scandium

TABLE I

The periodic table of elements showing the electron configuration
(simplified to show build-up of principal quantum numbers)

Element		Atomic number	
Hydrogen	H	1	$(1s)^1$ $n = 1, l = 0$
Helium	He	2	$(1s)^2$
Lithium	Li	3	,, $(2s)^1$ $n = 2, l = 0$
Beryllium	Be	4	,, $(2s)^2$
Boron	B	5	,, ,, $(2p)^1$ $n = 2, l = 1$
Carbon	C	6	,, ,, $(2p)^2$
Nitrogen	N	7	,, ,, $(2p)^3$
Oxygen	O	8	,, ,, $(2p)^4$
Fluorine	F	9	,, ,, $(2p)^5$
Neon	Ne	10	,, ,, $(2p)^6$
Sodium	Na	11	,, ,, ,, $(3s)^1$ $n = 3, l = 0$
Magnesium	Mg	12	,, ,, ,, $(3s)^2$
Aluminium	Al	13	,, ,, ,, ,, $(3p)^1$ $n = 3, l = 1$
Silicon	Si	14	,, ,, ,, ,, $(3p)^2$
Phosphorus	P	15	,, ,, ,, ,, $(3p)^3$
Sulphur	S	16	,, ,, ,, ,, $(3p)^4$
Chlorine	Cl	17	,, ,, ,, ,, $(3p)^5$
Argon	A	18	,, ,, ,, ,, $(3p)^6$
Potassium	K	19	,, ,, ,, ,, ,, $(4s)^1$ $n = 4, l = 0$
Calcium	Ca	20	,, ,, ,, ,, ,, $(4s)^2$

Element	Symbol	Z	Configuration	
Scandium	Sc	21	$(1s)^2(2s)^2(2p)^6(3s)^2(3p)^6(4s)^2(3d)^1$	$n = 3, l = 2$
Titanium	Ti	22	,, ,, ,, ,, ,, ,, $(3d)^2$	
Vanadium	V	23	,, ,, ,, ,, ,, ,, $(3d)^3$	
Chromium	Cr	24	,, ,, ,, ,, ,, ,, $(3d)^4$	
Manganese	Mn	25	,, ,, ,, ,, ,, ,, $(3d)^5$	
Iron	Fe	26	,, ,, ,, ,, ,, ,, $(3d)^6$	
Cobalt	Co	27	,, ,, ,, ,, ,, ,, $(3d)^7$	
Nickel	Ni	28	,, ,, ,, ,, ,, ,, $(3d)^8$	
Copper	Cu	29	,, ,, ,, ,, ,, ,, $(3d)^9$	
Zinc	Zn	30	,, ,, ,, ,, ,, ,, $(3d)^{10}$	
Gallium	Ga	31	,, ,, ,, ,, ,, ,, ,, $(4p)^1$	$n = 4, l = 1$
Germanium	Ge	32	,, ,, ,, ,, ,, ,, ,, $(4p)^2$	
Arsenic	As	33	,, ,, ,, ,, ,, ,, ,, $(4p)^3$	
Selenium	Se	34	,, ,, ,, ,, ,, ,, ,, $(4p)^4$	
Bromine	Br	35	,, ,, ,, ,, ,, ,, ,, $(4p)^5$	
Krypton	Kr	36	,, ,, ,, ,, ,, ,, ,, $(4p)^6$	
Rubidium	Rb	37	,, ,, ,, ,, ,, ,, ,, ,, $(5s)^1$	$n = 5, l = 0$
Strontium	Sr	38	,, ,, ,, ,, ,, ,, ,, ,, $(5s)^2$	
Yttrium	Y	39	,, ,, ,, ,, ,, ,, ,, ,, ,, $(4d)^1$	$n = 4, l = 2$
Zirconium	Zr	40	,, ,, ,, ,, ,, ,, ,, ,, ,, $(4d)^2$	
Niobium	Nb	41	,, ,, ,, ,, ,, ,, ,, ,, ,, $(4d)^3$	
Molybdenum	Mo	42	,, ,, ,, ,, ,, ,, ,, ,, ,, $(4d)^4$	
Technetium	Tc	43	,, ,, ,, ,, ,, ,, ,, ,, ,, $(4d)^5$	
Rubidium	Ru	44	,, ,, ,, ,, ,, ,, ,, ,, ,, $(4d)^6$	
Rhodium	Rh	45	,, ,, ,, ,, ,, ,, ,, ,, ,, $(4d)^7$	
Palladium	Pd	46	,, ,, ,, ,, ,, ,, ,, ,, ,, $(4d)^8$	
Silver	Ag	47	,, ,, ,, ,, ,, ,, ,, ,, ,, $(4d)^9$	
Cadmium	Cd	48	,, ,, ,, ,, ,, ,, ,, ,, ,, $(4d)^{10}$	

TABLE I (*continued*)

Element		Atomic number		
Indium	In	49	$(1s)^2(2s)^2(2p)^6(3s)^2(3p)^6(4s)^2(3d)^{10}(4p)^6(5s)^2(4d)^{10}(5p)^1$	$n = 5, l = 1$
Tin	Sn	50	,, ,, ,, ,, ,, ,, ,, ,, ,, ,, $(5p)^2$	
Antimony	Sb	51	,, ,, ,, ,, ,, ,, ,, ,, ,, ,, $(5p)^3$	
Tellurium	Te	52	,, ,, ,, ,, ,, ,, ,, ,, ,, ,, $(5p)^4$	
Iodine	I	53	,, ,, ,, ,, ,, ,, ,, ,, ,, ,, $(5p)^5$	
Xenon	Xe	54	,, ,, ,, ,, ,, ,, ,, ,, ,, ,, $(5p)^6$	
Caesium	Cs	55	,, ,, ,, ,, ,, ,, ,, ,, ,, ,, ,, $(6s)^1$	$n = 6, l = 0$
Barium	Ba	56	,, ,, ,, ,, ,, ,, ,, ,, ,, ,, ,, $(6s)^2$	
Lanthanum	La	57	,, ,, ,, ,, ,, ,, ,, ,, ,, ,, ,, ,, $(5d)^1$	$n = 5, l = 2$
Cerium	Ce	58	,, ,, ,, ,, ,, ,, ,, ,, ,, ,, ,, ,, ,, $(4f)^1$	$n = 4, l = 3$
Praseodymium	Pr	59	,, ,, ,, ,, ,, ,, ,, ,, ,, ,, ,, ,, ,, $(4f)^2$	
Neodymium	Nd	60	,, ,, ,, ,, ,, ,, ,, ,, ,, ,, ,, ,, ,, $(4f)^3$	
Promethium	Pm	61	,, ,, ,, ,, ,, ,, ,, ,, ,, ,, ,, ,, ,, $(4f)^4$	
Samarium	Sm	62	,, ,, ,, ,, ,, ,, ,, ,, ,, ,, ,, ,, ,, $(4f)^5$	
Europium	Eu	63	,, ,, ,, ,, ,, ,, ,, ,, ,, ,, ,, ,, ,, $(4f)^6$	
Gadolinium	Gd	64	,, ,, ,, ,, ,, ,, ,, ,, ,, ,, ,, ,, ,, $(4f)^7$	
Terbium	Tb	65	,, ,, ,, ,, ,, ,, ,, ,, ,, ,, ,, ,, ,, $(4f)^8$	
Dysprosium	Dy	66	,, ,, ,, ,, ,, ,, ,, ,, ,, ,, ,, ,, ,, $(4f)^9$	
Holmium	Ho	67	,, ,, ,, ,, ,, ,, ,, ,, ,, ,, ,, ,, ,, $(4f)^{10}$	
Erbium	Er	68	,, ,, ,, ,, ,, ,, ,, ,, ,, ,, ,, ,, ,, $(4f)^{11}$	
Thulium	Tm	69	,, ,, ,, ,, ,, ,, ,, ,, ,, ,, ,, ,, ,, $(4f)^{12}$	
Ytterbium	Yb	70	,, ,, ,, ,, ,, ,, ,, ,, ,, ,, ,, ,, ,, $(4f)^{13}$	
Lutetium	Lu	71	,, ,, ,, ,, ,, ,, ,, ,, ,, ,, ,, ,, ,, $(4f)^{14}$	
Hafnium	Hf	72	,, ,, ,, ,, ,, ,, ,, ,, ,, ,, ,, ,, ,, $(5d)^2$	$n = 5, l = 2$
Tantalum	Ta	73	,, ,, ,, ,, ,, ,, ,, ,, ,, ,, ,, ,, ,, $(5d)^3$	
Tungsten	W	74	$(5d)^4$	

Osmium	Os	76	"	"	"	"	"	"	"	"	"	"	"	"	"	$(5d)^6$					
Iridium	Ir	77	"	"	"	"	"	"	"	"	"	"	"	"	"	$(5d)^7$					
Platinum	Pt	78	"	"	"	"	"	"	"	"	"	"	"	"	"	$(5d)^8$					
Gold	Au	79	"	"	"	"	"	"	"	"	"	"	"	"	"	$(5d)^9$					
Mercury	Hg	80	"	"	"	"	"	"	"	"	"	"	"	"	"	$(5d)^{10}$					
Thallium	Tl	81	"	"	"	"	"	"	"	"	"	"	"	"	"	"	$(6p)^1$				$n = 6, l = 1$
Lead	Pb	82	"	"	"	"	"	"	"	"	"	"	"	"	"	"	$(6p)^2$				
Bismuth	Bi	83	"	"	"	"	"	"	"	"	"	"	"	"	"	"	$(6p)^3$				
Polonium	Po	84	"	"	"	"	"	"	"	"	"	"	"	"	"	"	$(6p)^4$				
Astatine	At	85	"	"	"	"	"	"	"	"	"	"	"	"	"	"	$(6p)^5$				
Radon	Rn	86	"	"	"	"	"	"	"	"	"	"	"	"	"	"	$(6p)^6$				
Francium	Fr	87	"	"	"	"	"	"	"	"	"	"	"	"	"	"	"	$(7s)^1$			$n = 7, l = 0$
Radium	Ra	88	"	"	"	"	"	"	"	"	"	"	"	"	"	"	"	$(7s)^2$			
Actinium	Ac	89	"	"	"	"	"	"	"	"	"	"	"	"	"	"	"	"	$(6d)^1$		$n = 6, l = 2$
Thorium	Th	90	"	"	"	"	"	"	"	"	"	"	"	"	"	"	"	"	"	$(5f)^1$	$n = 5, l = 3$
Protoactinium	Pa	91	"	"	"	"	"	"	"	"	"	"	"	"	"	"	"	"	"	$(5f)^2$	
Uranium	U	92	"	"	"	"	"	"	"	"	"	"	"	"	"	"	"	"	"	$(5f)^3$	
Neptunium	Np	93	"	"	"	"	"	"	"	"	"	"	"	"	"	"	"	"	"	$(5f)^4$	
Plutonium	Pu	94	"	"	"	"	"	"	"	"	"	"	"	"	"	"	"	"	"	$(5f)^5$	
Americium	Am	95	"	"	"	"	"	"	"	"	"	"	"	"	"	"	"	"	"	$(5f)^6$	
Curium	Cm	96	"	"	"	"	"	"	"	"	"	"	"	"	"	"	"	"	"	$(5f)^7$	
Berkelium	Bk	97	"	"	"	"	"	"	"	"	"	"	"	"	"	"	"	"	"	$(5f)^8$	
Californium	Cf	98	"	"	"	"	"	"	"	"	"	"	"	"	"	"	"	"	"	$(5f)^9$	
Einsteinium	Es	99	"	"	"	"	"	"	"	"	"	"	"	"	"	"	"	"	"	$(5f)^{10}$	
Fermium	Fm	100	"	"	"	"	"	"	"	"	"	"	"	"	"	"	"	"	"	$(5f)^{11}$	
Mendeleevium	Md	101	"	"	"	"	"	"	"	"	"	"	"	"	"	"	"	"	"	$(5f)^{12}$	
Nobelium	No	102	"	"	"	"	"	"	"	"	"	"	"	"	"	"	"	"	"	$(5f)^{13}$	
Lawrencium	Lr	103	"	"	"	"	"	"	"	"	"	"	"	"	"	"	"	"	"	$(5f)^{14}$	

$(3d)^1$ and yttrium $(4d)^1$ for the two previous periods. The electron energy levels are now so high that the difference between the various quantum number configurations is decreasing rapidly as the levels recede from the nucleus. Consequently, $6s$, $5d$, $4f$, and $6p$ are very similar indeed. Element number 58, cerium, does not follow lanthanum by filling up the $5d$ electron level, but instead the additional electron reverts to an energy level of lower quantum state, the $4f$, where $n = 4$ and $l = 3$. This can occur solely because of the close proximity of the energy levels mentioned, and in fact gives rise to a lower state of free energy in the cerium atom than would be the case if this element contained two $5d$ electrons. A new group of elements then begins, in which the $4f$ electrons fill up to the maximum number, namely $2 \times 7 = 14$ ($m_l = +3$ to -3). This group is inserted immediately after lanthanum, and is often referred to as the *lanthanide group*. Since the $4f$ electrons are not the farthest away from the nucleus, the chemistry of the lanthanide elements is very similar, and separation of one element from the other can be a laborious process. This fact accounts for their original name—the 'rare earths'. They are all extremely reactive, and difficult to produce in the elemental form because of the rapid formation of an oxide film on the metal—hence the term 'earth', meaning oxide. The last of the group, lutetium ($Z = 71$), completes the $4f$ energy level $(4f)^{14}$ and the next element, hafnium, reverts to the process commenced at lanthanum, namely the filling of the $5d$ level, while retaining, of course, the fourteen $4f$ electrons. Thereafter the $5d$ level progresses to its maximum of ten electrons, elements in this period resembling those with similar numbers of $4d$ or $3d$ electrons in the two previous long periods. Mercury ($Z = 80$) completes this process, and then thallium ($Z = 81$) commences a new p group, the $6p$ state. Once again the elements whose highest energy level electrons are in $6p$ have somewhat unusual properties, some being only semimetallic. The six $6p$ states and the fifth period are completed with the rare gas radon ($Z = 86$).

The last period of the system continues the familiar process of building up first to two $7s$ electrons (radium), then commencing with $6d$ as lanthanum did with $5d$. The element concerned is actinium ($Z = 89$) and, like lanthanum, it is followed by a new group, similar to the lanthanides, but having electrons this time in the $5f$ level. The first of these is thorium ($Z = 90$) which strongly resembles cerium, then protoactinium ($Z = 91$) and uranium ($Z = 92$). This was the last of the known elements before the discovery of nuclear fission, but the whole group can now be completed with the isolation of every element up to $Z = 103$, lawrencium. These are called the 'actinide' elements, for the same reasons that the rare earths are

B

TABLE II

GROUPS

BELOW FULL AND DOTTED LINES: —— METALS
BETWEEN FULL AND DOTTED LINES: —— SEMI-METALS
ABOVE FULL LINE: —— NON-METALS

	IA	IIA		IIIA	IVA	VA	VIA	VIIA	VIII	VIII	VIII	IB	IIB		IIIB	IVB	VB	VIB	VIIB	
1s	1 H																			2 He
2s	3 Li	4 Be												2p	5 B	6 C	7 N	8 O	9 F	10 Ne
3s	11 Na	12 Mg												3p	13 Al	14 Si	15 P	16 S	17 Cl	18 A
4s	19 K	20 Ca	3d	21 Sc	22 Ti	23 V	24 Cr	25 Mn	26 Fe	27 Co	28 Ni	29 Cu	30 Zn	4p	31 Ga	32 Ge	33 As	34 Se	35 Br	36 Kr
5s	37 Rb	38 Sr	4d	39 Y	40 Zr	41 Nb	42 Mo	43 Tc	44 Ru	45 Rh	46 Pd	47 Ag	48 Cd	5p	49 In	50 Sn	51 Sb	52 Te	53 I	54 Xe
6s	55 Cs	56 Ba	5d	57 La	72 Hf	73 Ta	74 W	75 Re	76 Os	77 Ir	78 Pt	79 Au	80 Hg	6p	81 Tl	82 Pb	83 Bi	84 Po	85 At	86 Rn
7s	87 Fr	88 Ra	6d	89 Ac																

4f	58 Ce	59 Pr	60 Nd	61 Pm	62 Sm	63 Eu	64 Gd	65 Tb	66 Dy	67 Ho	68 Er	69 Tm	70 Yb	71 Lu	IIIA
5f	90 Th	91 Pa	92 U	93 Np	94 Pu	95 Am	96 Cm	97 Bk	98 Cf	99 Es	100 Fm	101 Md	102 No	103 Lr	IIIA

ELECTRON LEVEL DETERMINING VALENCY AND MAIN PROPERTIES

called lanthanides, but only the first three actinides are natural, and all the remainder are 'transuranic' elements. There is still some doubt about the precise electron configuration here, e.g. it has been suggested that thorium has no $5f$ electron and ought not to be counted as an actinide. Since the chemistry of each is comparable, thorium may be best left to fit into the same periodic group as cerium.

The table of electron configurations that can be evolved from the above arguments (Table I) cannot indicate at the same time the chemical resemblance of particular elements. Table II, showing the same classification but in related groups and periods, identifies the elements in terms of their related main properties. Table II shows only the electron groups of highest energy level, or alternatively the characteristic energy levels that are filling up to complete a group such as the transition, lanthanide, or actinide series, since here the levels concerned are not those of the highest energy. Table II also summarizes the division of elements, referred to previously, into non-metals, semimetals, and true metals. The relevance of the p electron state will be obvious. This type of division will be enlarged on in Chapter 2 in terms of the bound atom. The discussion so far has concerned only the 'free' atom, i.e. the atom in the vapour state, or present in the form of an 'ion', having lost one or more electrons to infinity. In the solid state, with which materials science is more concerned, there are modifications to the electron pattern. Nevertheless, the structure of the free atom is of immense importance in appreciating the classification of elements, their chemistry, and their possible relationships with one another in combination or in solution. These electron states, derived from the four quantum numbers, have explained many anomalies in the periodic system.

1.6. *Isotopes*

One further factor affecting the application of materials can be considered at this point: the effect on the atom of changes in the nucleus. The nucleus consists primarily of *protons*, positively charged particles of much greater mass than the electrons (1,870 times) but of equal number to the extra-nuclear electrons, and *neutrons* of mass equal to that of the proton but without charge. The mass of an atom is therefore concentrated mainly in the nucleus, together with the positive charge, and the electron cloud contains the balancing negative charge with a much smaller total mass. When neutrons (uncharged particles of mass equal to that of the proton) are added to the nucleus, the only disturbance to this balance is one of gravitational forces. The electrons are hardly affected, particularly those remote from the nucleus in an element of high atomic number. The latter are shielded from the nucleus by those at lower energy levels, and the higher energy levels are most concerned in chemical and other properties. Obviously, little change in general properties occurs when neutrons are added or subtracted from the nucleus, and only those involving mass effects would be noticed. The same atom with more or less neutrons thus behaves in an almost identical manner as originally, but is heavier or lighter. Such atoms are referred to as *isotopes*.

Natural elements commonly contain a mixture of these isotopes, and the atomic weight is therefore a mean value dependent on the particular isotopes and their proportion. Few neutrons exist in the elements of lower atomic number, although hydrogen has three isotopes, the two heavier ones being known as deuterium (originally called heavy hydrogen), with one neutron per atom, and tritium, with two neutrons per atom. Tritium is unstable. The elements of higher atomic number contain at least as many neutrons as protons, but because of the shielding effect of the innermost electrons this does not affect the material properties. In the atoms of very high atomic number, the number of neutrons considerably exceeds the number of protons, and these elements are often referred to as the *heavy elements*. In uranium atoms, for instance, there are 92 electrons and therefore 92 protons. The most common isotope has, in addition, 146 neutrons, so that its atomic weight is 238. We can write the chemical symbol as $^{238}_{92}U$ to indicate the atomic number and weight of this particular isotope. Natural uranium contains 99% of this isotope. Other significant isotopes of uranium include $^{235}_{92}U$, which occurs to the extent of 0·7% in natural uranium, and $^{233}_{92}U$ which is artificially produced by neutron bombardment of thorium.

The nucleus itself eventually becomes unstable, and the addition of a further neutron may cause nuclear 'fission', as with ^{235}U and ^{233}U. The atom then disintegrates, with the production of two new atoms and a debris of elementary particles and radiations. Certain elements are so unstable that neutron bombardment is unnecessary: there is a spontaneous disintegration. These are the naturally radioactive isotopes, so called because radiation is invariably emitted during their decay. This decay may be very slow, but sometimes it is remarkably rapid. Almost all elements of atomic number greater than $Z = 83$ (bismuth) are naturally radioactive. Polonium ($Z = 84$) contains 126 neutrons, representing a neutron/proton ratio of 1·5. In fact there are only two elements with a lower atomic number and lower neutron-to-proton ratio than this which are radioactive.

They are technetium ($Z = 43$), with 56 neutrons, and promethium ($Z = 61$), one of the rare earths, with 86 neutrons in its most common isotope. The transuranic elements produced by neutron adsorption in atoms of ^{238}U or its isotopes, and subsequent radioactive decay, are mostly very short-lived, and many undergo fission.

2

The Bound Atom

2.1. *Band structure*

The 'free' atoms discussed in Chapter 1 do exist, under appropriate conditions, but the vast majority of engineering materials are handled in the liquid or solid state, particularly in the solid. The electrons, and the protons and neutrons in the nucleus of an atom, are not altered simply by the condensation into solid or liquid, since any variation in the number of these elementary particles must imply a change in constitution; either the production of a new isotope if only neutrons are involved, or transmutation to a new element if protons and therefore electrons are concerned. The close approach of two or more atoms of the same element does, however, affect the *energy* distribution of the electrons, and therefore has a marked influence on the properties.

The discrete energy levels of electrons in the free atom are fairly well separated close to the nucleus, but as explained in Chapter 1 (page 14) the levels are much closer as they recede from the nucleus, and above 4*s*, 4*p*, etc., there is little to choose between them. When two atoms of like kind are brought together, the outer electrons, in the higher energy levels, will therefore be closest. The inner electrons will be partly screened by these outer electrons, as is the nucleus, from any influence of the other atom. As the outer electron clouds become very close, the distinction between some of these higher energy levels becomes rather diffuse. The net effect is to broaden the discrete energy levels into 'bands', so that a 4*s* electron, for instance, may now exist over a broader range of energy without changing its state. The closer the atoms approach, the greater the broadening, and the deeper into the atom are electrons affected.

By plotting the energy levels (for a relatively simple atom) against the distance apart of two atoms, this effect is fairly obvious. Figure 2.1 shows the energy levels in magnesium, as the atoms come together. It will be obvious that electrons at the 1*s* level are not affected until the distance apart of the atoms is very small indeed. The 2*s* and 2*p* levels are affected before the 1*s*, and, of course, the 3*s* earlier still. Although there are no electrons normally in the 3*p* state in magnesium, the broadening still occurs, for exactly the same reasons as it does for the normal occupied states. Hence, at some distance apart of the atoms, overlapping of the 3*s* and 3*p* bands will inevitably occur, and the normal 3*s* electrons of magnesium are now no longer restricted to that level but can occupy the 3*p* state also, *without* excitation. In other words, there is now a wider band of energy levels available to the electrons, not only in the normal ground state of, in this example, 3*s*, but also by virtue of the 'smearing' of this band into the 3*p* band. The two outer electrons of magnesium can therefore occupy a 'hybrid' type of energy level, somewhere between the lower limit of the 3*s* band and the upper limit of the 3*p* band. It becomes difficult to decide exactly where these electrons are in such an example, but fortunately it is not always essential to know this precisely, provided that it is recognized that the electrons may exist freely between the bounds of the 3*s* and 3*p* levels, broadened out as they now are by this approach of two atoms. Were the atoms to continue their approach, the 2*s* and 2*p* levels might also begin to overlap, as Figure 2.1 shows. There is, however, an optimum distance of approach, because the mutual repulsion of the electron clouds begins to take effect. The optimum distance is that taken up when the atoms are condensed into the solid state, usually the crystalline form of that solid, and depends on temperature. In the example of Figure 2.1, the point X shows this optimum distance and the dotted line illustrates the broadening of the 3*s* and 3*p* level in magnesium at this distance. This broadening of energy levels into bands takes place, to a greater or lesser degree, in all atoms when condensed, and has very important influences on the physical properties, particularly conductivity. This point will be dealt with later in this chapter. The broadening of energy levels gives rise to what is termed the 'valence' state of the electronic configurations, since it refers to bound atoms as distinct from free ones. The change from the free electron state to the valence state is obviously highly significant, but it must be remembered that the inner electrons are hardly affected, and the electron configuration of the free atom must be borne in mind in determining the subsequent effect of binding the atoms together.

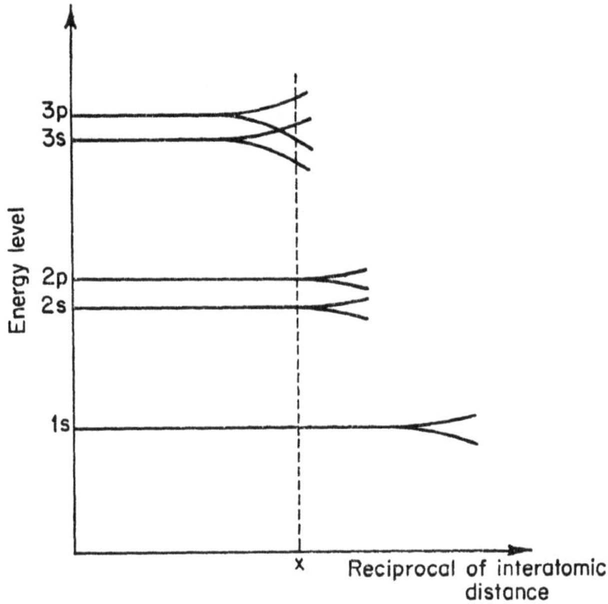

FIG. 2.1. Energy levels of electron states in magnesium.

2.2. *The attractive and repulsive forces in binding*

The free atom is stable, with the electrostatic attraction between the electron cloud and the positively charged nucleus balanced by the gravitational forces. When two similar atoms are brought together, there is initially no change in either atom with regard to this balance of forces. Eventually a point is reached at which the nucleus of each atom attracts the electron cloud of the other, and this attraction increases as the distance apart decreases. There are other attractive forces involved which depend on the type of bond. These will be referred to later, but for the present they can be regarded as additional forces of attraction to supplement the simple electrostatic one referred to.

As the attractive force increases, the atoms draw closer together and the effect on the energy levels, broadening these into bands, becomes more noticeable. The electron clouds are now almost overlapping and forces of repulsion begin to build up, primarily between the electron clouds. Once more, there are also other forces of repulsion that must be added to the simple electrostatic one; these will be dealt with under the appropriate headings. A balance between repulsive and attractive forces is reached, at which point the assembly reaches the minimum potential energy, the resultant force is zero, and the assembly is stable. Closer approach of the atoms causes a rise in potential energy until it becomes positive, and the assembly would be unstable again. Atoms cannot therefore approach closer than this particular interatomic distance, which is referred to simply as the closest distance of approach.

Figure 2.2 shows the plot of potential energy (from all sources

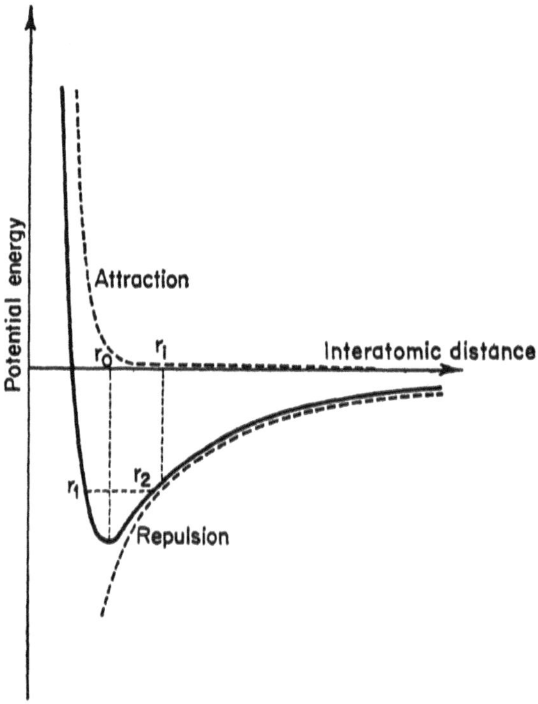

FIG. 2.2. Potential energy of attraction and repulsion between atoms.

both repulsive and attractive), against the interatomic distance. The upper (broken) curve represents the repulsion energy and the lower (broken) curve the energy of attraction. The full curve gives the resultant and illustrates the minimum potential energy at which stability is reached. The total interaction energy is negative at this point, otherwise the atoms would not be bound together, and both attractive and repulsive energy are inversely proportional to some power of interatomic distance. The attraction energy is negative, since the atoms do the work of attraction. The potential energy of attraction is equal to $-a/r^m$ and the potential energy of repulsion is $+b/r^n$, where a and b are proportionality constants, and n and m are powers of r, the interatomic distance.

The net potential energy is equal to

$$-\frac{a}{r^m}+\frac{b}{r^n} \tag{2.1}$$

This equation, due to MIE, represents that of the full curve in Figure 2.2. The evaluation of the constants a and b, and particularly the indices m and n, is not a simple matter. As stated previously, the potential energy of attraction and repulsion can be the summation of several effects acting simultaneously. Nevertheless, several aspects of the properties of materials and the electronic structure of some solid substances can be illustrated by means of this equation, and it will be referred to again, in Chapters 5 and 7 in particular.

The two factors so far discussed, the band structure of the electrons and the balance of attraction and repulsion in the bound atom, can now be amplified in the light of particular applications, and the types of bonding that result in certain specific classes of materials.

2.3. *Stable electron configurations*

In Chapter 1, dealing with the free atom, reference was made to the comparative chemical inertness of those elements that complete a particular quantum group in the periodic system, i.e. the rare gases. All these rare gases have a completed group or subgroup of 2 (helium), 8 (neon, argon), 18 (krypton, xenon), or 32 (radon) electrons in their respective periods. Each of them completes a subgroup, i.e. a particular value of the quantum numbers n and l, at the end of the period, and an increase of one occurs, for the element of next higher atomic number, in the value of n. This led to the conclusion that greater stability results from completion of these electron groups and subgroups than from any other cause. Moreover, the inert gases have very low boiling points, i.e. they are normally gases, and therefore free atoms, over a very wide range of temperature, whereas most other materials tend to bind together as solids or liquids. Those that are gases but not in the inert gas group nevertheless have higher boiling points and are much more reactive.

Some fifty years or so ago it was suggested that other substances might achieve a measure of stability in forming molecules between like atoms, or chemical compounds between unlike atoms, if the inert gas stable configuration were approached by electrons moving from one atom to another to complete the stable subgroup. This stable subgroup was taken to be the octet of electrons, the maximum for a given combination of n and l. (Helium is an exception with 2, but still stable.) Taking the outer electrons only (for detail see Table I) we have:

Helium	$(1s)^2$
Neon	$(2s)^2(2p)^6$
Argon	$(3s)^2(3p)^6$
Krypton	$(4s)^2(4p)^6$
Xenon	$(5s)^2(5p)^6$
Radon	$(6s)^2(6p)^6$

All other atoms must have more electrons than the rare gas preceding them in the periodic system, and less than the succeeding one. If an atom can lose the excess electrons over those of the preceding rare gas, it retains the stable octet of that rare gas as its outer electrons. If, on the other hand, it gains the number of electrons to make up to the number in the succeeding rare gas, it also achieves a stable octet. Such processes are obviously limited, yet they account for two main classes of chemical compounds: (i) those in which electrons are donated by one kind of atom and accepted by another, and (ii) those in which electrons are continuously shared between atoms. These two groups will be dealt with separately.

2.4. *Ionic bonding*

The classic case of this type of bonding is that of sodium chloride NaCl, common salt. If the electronic structures of the rare gases closest to sodium and chlorine are written down in juxtaposition to those of the two elements concerned we have:

Neon	$(1s)^2(2s)^2(2p)^6$
Sodium	$(1s)^2(2s)^2(2p)^6(3s)^1$
Chlorine	$(1s)^2(2s)^2(2p)^6(3s)^2(3p)^5$
Argon	$(1s)^2(2s)^2(2p)^6(3s)^2(3p)^6$

It is obvious that sodium has one more electron than neon, and seven less than argon, whereas chlorine has one less than argon and seven more than neon. The simplest way to attain the stable octet is for sodium to give up its one $3s$ electron (so that sodium now has the stable octet of neon) and for chlorine to accept this $3s$ electron to make up its $3p$ group to six, whereby chlorine assumes the stable octet of argon. In such a process, the sodium atom is now short of one electron, and so has a net positive charge. Similarly, the chlorine atom now has one more electron than the number of protons in its nucleus, and is negatively charged. Normally this state of affairs could not prevail for long, were it not for the obvious and immediate electrostatic attraction that now results between the positive sodium ion and the negative chlorine one. Both atoms retain the stable octet that resulted from the donation and acceptance of one electron between each pair of atoms, but they are strongly attracted. As they draw closer together, the electron clouds will begin to overlap and to undergo the repulsion that occurs when any two atoms approach closely. Here, however, we have an extraordinary force of attraction, involving not merely the nucleus and electron clouds, but also the imbalance of charges that has resulted from the formation of two oppositely charged ions. The forces of attraction are therefore very high. The potential energy of attraction (giving rise to so-called 'Coulomb forces') is given by

$$\phi_C = \frac{-e_1e_2}{r} \tag{2.2}$$

where e_1 and e_2 are the charges involved and r is the interatomic distance.

With more complex molecules, when one or more of the ions may be a dipole, or carry a more complex charge, then the potential energy is modified by the non-uniformity of the charges. The effect of polarization is to modify the power of r in the denominator of equation (2.2).

When the electron clouds overlap, however, there will be the customary broadening of energy levels discussed in Section 2.1. In addition, the strong forces of attraction will cause the relevant energy levels of the two ions to become a common system, e.g. the $(2p)^6$ electrons from the sodium ion and the $(2p)^6$ electrons from the chlorine ion may both be overlapping in the molecule. The Pauli exclusion principle does not, however, allow more than one electron to occupy the same energy level as defined by all four quantum numbers so that some of these electrons must move into a different level. The broadening of the energy levels into bands does not alleviate the position when the maximum number of electrons is present in any band, which is precisely the case in ionic bonding, since, by definition, this is a state of stable octets, i.e. every energy level is completely filled. It is therefore necessary for some electrons to move into a higher level, to observe the Pauli principle. The $(2p)^6$ electrons from one ion must move into a higher level. The $(3s)$ and $(3p)$ electron subgroups are also filled, however, by the stable octet of the chlorine ion. There must be some movement, therefore, into even higher levels. The net effect of this must be an increase in energy of the whole molecule, and positive energy here means repulsion (see Figure 2.2). This repulsion counteracts the increased Coulomb attraction forces and so a balance is restored and stability assured. In *ionic* compounds, therefore, both attraction and repulsion are increased, the first by the type of bond increasing electrostatic (Coulomb) forces, and the second as a result of the *exchange* repulsion due to the complete filling of electron energy levels, even when the latter are broadened into bands.

The resultant bond is still stronger than the simple case discussed in Section 2.2, by virtue of the increased net attraction. Not only is the mechanical strength relatively high, but also the melting point. The argument so far has referred to two atoms forming one molecule, but in the solid state these molecules build up into a crystal lattice. The ionic charge is distributed in a spherical manner around each atom, and the atoms pack closely together in the crystal according to the relative sizes of each ion and the charge on each. The sum of electric charges must be zero, so that the ionic solid is electrically neutral. The ions therefore arrange themselves in such a manner as to achieve neutrality and obtain the closest possible packing of spheres. In sodium chloride, the chlorine ion is larger than the sodium, and the crystal structure approximates to that of a simple cube, at the corners of which are, alternately, sodium and chlorine atoms, the cubic packing extending in this manner along three

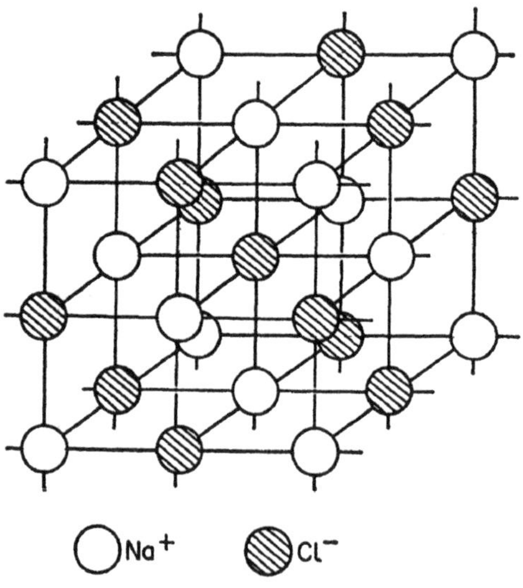

FIG. 2.3. The structure of sodium chloride.

mutually perpendicular directions. The cube side is fixed by the larger chlorine atom, with the sodium atoms occupying the spaces between. Figure 2.3 illustrates the basic cubic lattice, and Plate III the packing of atoms in their relative sizes in NaCl.

Where the atoms in ionic compounds contain more than one electron above the stable configuration, then more than one atom of the donor or acceptor type may be involved. Magnesium, for instance, will form an ionic compound with two chlorine atoms, $MgCl_2$, by donating its two 3*s* electrons, one to each of two chlorine atoms. The magnesium atom will be doubly charged (positive) and will therefore take up a position in the crystal lattice in which each magnesium atom is close to two chlorine atoms, so as to preserve the neutrality.

2.5. *Covalent bonding*

The second method by which atoms can achieve the stable octet is by sharing electrons. When this occurs, the shared electrons cannot be regarded as 'belonging' to one or other atoms forming the molecule, but to the shared group; inevitably the properties of such a bond must be different from that of ionic compounds. Hydrogen atoms have one electron, and the nearest inert gas configuration is that of helium.

Hydrogen	$(1s)^1$
Helium	$(1s)^2$

If two hydrogen atoms share their electrons, then each can be regarded as having, momentarily at least, a stable configuration. To satisfy the Pauli principle these two shared electrons must have opposite spin, as they do in helium. The hydrogen molecule H_2 possesses much more stability than individual atoms of hydrogen.

Similarly fluorine atoms, one short of the stable configuration of neon, can form a shared pair.

Fluorine	$(1s)^2(2s)^2(2p)^5$
Neon	$(1s)^2(2s)^2(2p)^6$

Two fluorine atoms can share the two electrons donated one by each, so completing the octet. Figure 2.4 shows, diagrammatically,

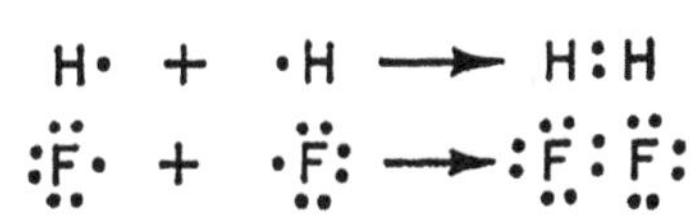

FIG. 2.4. The covalent bond in hydrogen and fluorine.

the outer electron distribution in both hydrogen and fluorine molecules. Again, fluorine molecules F_2 will be more stable than individual atoms, and again, the shared electrons must be of opposite spin to satisfy the Pauli principle.

In both these cases, only one pair of electrons is shared, and this is a single bond. The type of bonding is known as *covalent* or *homopolar*, the latter term serving to emphasize the difference from the ionic bond, which is polarized by virtue of the ionic charges.

When the atom concerned is short of more than one electron to complete a stable octet, then more than one bond may be involved. Double covalent bonds involve four shared electrons in two pairs of opposite spin. Triple bonds will naturally require six shared electrons in three pairs of opposite spin.

Oxygen atoms are two electrons short of the stable octet, and therefore form double bonds by sharing four electrons in the oxygen molecule O_2 (two from each atom). Nitrogen atoms require three electrons to make up the octet, and so produce a triple bond with six shared electrons in the molecule N_2 (three from each atom). Because of the requirement for pairs of electrons, of opposite spin, covalent bonds are directional, unlike the spherical distribution of the charges in ionic compounds. Thus, with nitrogen, the bonds in its covalent compounds, such as ammonia NH_3, take up three

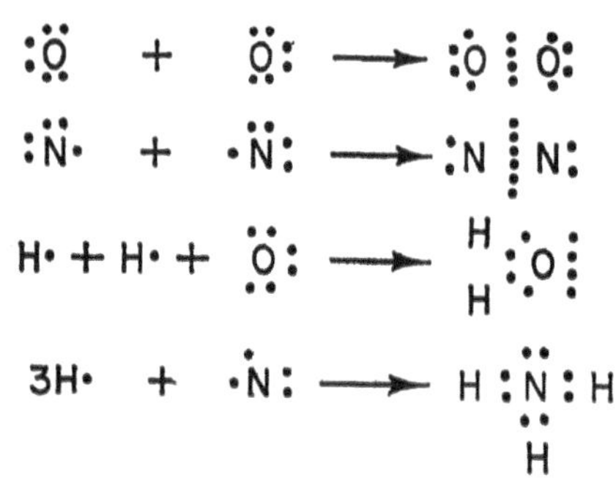

FIG. 2.5. The covalent bond in oxygen, nitrogen, water, and ammonia.

mutually perpendicular directions. Figure 2·5 illustrates the molecules of oxygen O, nitrogen N, water H_2O, and ammonia NH_3, with their shared electrons, shown in two dimensions only.

The classical example of covalent bonding occurs in the chemistry of carbon and its compounds. Carbon has four electrons short of the stable octet.

$$C \quad (1s)^2(2s)^2(2p)^2$$

The sharing of four electrons from other atoms would therefore allow covalent bonding. To satisfy the Pauli principle, we must pair off the shared electrons to give opposite spin pairs. If this happens only to the four shared electrons required by carbon, then there will be a double bond of two opposite spin pairs. The existing two $2p$ electrons will also have to change to an opposite instead of parallel spin pair, the two $2s$ electrons being already an opposite spin pair. In the formation of solid carbon, however, the approach of other

FIG. 2.6. The structure of diamond.

carbon atoms causes a hybridization of one $2s$ electron, which moves into the $2p$ state, with an obvious increase in energy. The valence state of carbon is then actually $(1s)^2(2s)^1(2p)^3$ and the three $2p$ electrons and one $2s$ are of parallel spin. The four shared electrons required for covalency can now pair off with these four parallel-spin electrons already present, one in $2s$ and three in $2p$. Thus we can have four paired bonds rather than two formed from the shared electrons, and the higher energy of the $2s$ electron moving into a hybrid position is compensated by the lower binding energy that results from a four paired bond as compared with two in the simpler case. The hybrid state is written sp^3 to indicate the movement of one s electron into a p state. Once again, the four sp^3 bonds are highly directional, and take up a position directed towards the four corners of a regular tetrahedron. This is the structure of solid carbon in the allotropic form known as diamond. Diamond has a very high mechanical strength, and a high melting point. (The oxidation of diamond when heated in air must not be confused with true melting.) The covalent bond may be thus regarded as mechanically strong. Figure 2.6 illustrates the structure of diamond, and shows that each carbon atom has four close neighbours.

2.6. *Van der Waals' forces*

There are certain types of molecule in which a weak type of bonding exists, and where ionic or covalent bonding is not possible. Sometimes this type of bonding is present as part of the crystal structure, acting in particular directions or between certain groups of atoms.

An example exists with the rare gases. These have a complete octet of electrons in the outer levels, and therefore cannot bind together by ionic or covalent types of bond. They do, however, form molecules under certain conditions. Just as atoms without stable octets of electrons exercise an attraction between nucleus and electron cloud when

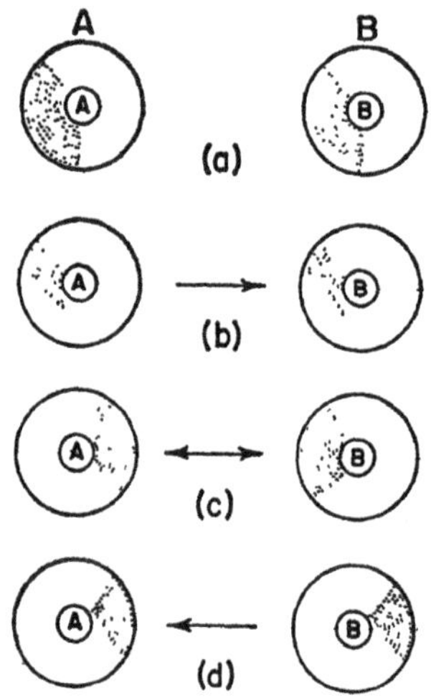

FIG. 2.7. Van der Waals' attraction between like atoms.

the atoms come close together, so the electron cloud, being continually moving, can exercise polarization forces over other atoms. At a particular moment in time, two approaching atoms of an inert gas such as argon can be envisaged (Figure 2.7a) such that one electron cloud tends to be concentrated on the side of the atom diametrically opposite to the close distance of approach, and that of the other on the same side as that of the approach. In this case, the nucleus of one atom A will be strongly attracted to the electron cloud of the other atom B (Figure 2.7b). The nucleus A is momentarily unshielded by its own electrons, and so moves closer to atom B. An instant later, the position may be changed (Figure 2.7c), and both electron clouds shield the two nuclei. Attraction ceases momentarily. At the next point in time (Figure 2.7d), the electron cloud of A is in a position to attract the nucleus of B whose own cloud is now on the wrong side. Such a position might be continuously changing, and a net attraction results, with a weak force of binding owing to the absence of other forces to overcome the normal repulsion of close approach. These forces are called *polarization* or *Van der Waals'* forces, after the originator who postulated their existence in explanation of the kinetic behaviour of gases. This type of bonding gives rise to molecular formations in the rare gases.

When the atoms or molecules are already polarized, they may 'induce' dipole moments in other atoms or molecules, giving rise to a further Van der Waals' attraction. The more complex the dipole moment of the two constituents, the more complex the attraction. The potential energy of Van der Waals' attraction, in the simplest case, is given by

$$\phi_{\text{v.d.w.}} = \frac{-3}{4}\frac{\alpha^2 E_0}{r^6} \tag{2.3}$$

where α represents the polarizibility, E_0 an energy value characteristic of the atom concerned, and r is the interatomic distance.

Compared with the potential energy of Coulomb attraction in ionic compounds (dependent on $1/r$) the Van der Waals' energy is obviously low. In other cases, however, Van der Waals' attraction assists the binding of certain types of crystal to give rise to unusual properties, as for example in graphite, and some semi-metals, such as arsenic, antimony, etc. These will be discussed in more detail in Chapter 3.

2.7. *The metallic bond*

In the solid state most metals are strongly bonded, which accounts for some of their useful applications. The metals fall into three main groups (see Table II), as follows:

- Alkali and alkaline-earth metals (e.g. sodium, magnesium) having outer *s* electrons.
- Transition metals having *d* or *f* electrons in outer subgroups (e.g. iron).
- Amphoteric and 'semi'-metals (e.g. aluminium and antimony) having *p* electrons, in the outer subgroup.

The metallic bond is a complex one, and varies somewhat between these three main groups. The type of crystal lattice adopted is also partly dependent on the outer electron groups, and there is also a trend towards a partial covalent bond in the third group.

Commencing with the simplest explanation of metallic bonding, which is certainly characteristic of sodium and the other alkali metals, it is necessary to invoke the band-structure argument of Section 2.1. As two sodium atoms approach, the broadening of the $3s$ energy level and its overlapping with a broadened empty $3p$ level allows considerable freedom to the $(3s)^1$ electron contributed by each atom, without any violation of the Pauli principle. The two $3s$ electrons merely form an electron pair with opposite spin, rather like the covalent bond, but *not* involving sharing of electrons to complete a stable configuration. If a third sodium atom now arrives, it cannot contribute its one $3s$ electron to the overlapped $3s$ state of the other two, because only two $3s$ electrons are allowed by the Pauli principle. The overlapping of the $3s$ and $3p$ broadened bands, however, allows this third $3s$ electron to enter quite readily into the hybridized band. The $3p$ state can accept a total of six electrons (three pairs of opposite

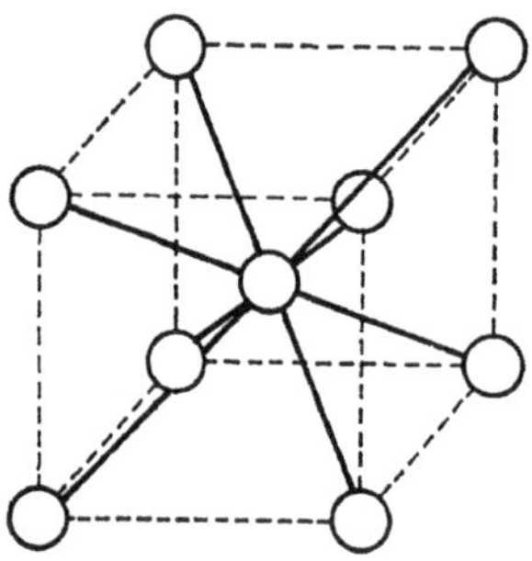

FIG. 2.8. The body-centred cube—sodium.

spin) and so this overlapped hybridized band could accept a total of eight electrons, two from the $3s$ and six from the $3p$. This means that eight sodium atoms could join the overlap of atoms and still satisfy the Pauli principle of the quantum theory, and a large bound group can be formed. In practice, the physical size of the atoms will also play a part, as it may not be possible to pack eight atoms close together in this way. The sodium atom can do so, and so forms a crystal lattice in which each atom is surrounded by eight neighbours. This type of structure will be described in more detail in Chapter 3 as the body-centred cube. An atom in the centre of a cubical box, with atoms at each corner of the box, is in fact surrounded by eight equidistant neighbours—hence the term body-centred cube (see Figure 2.8).

Since the sodium atom at the centre of the cube possesses only one $3s$ electron, and is in fact forming an opposite spin pair with one other atom, there must be a closer association with that second atom, as mentioned above, rather like a covalent bond shared electron pair. However, there is no stable octet or similar grouping, and to retain an equal bond with the other atoms in the cube, it was suggested by PAULING (not to be confused with Pauli) that the bond resonates between all eight of the near neighbours. Thus, Pauling sees the unit cell of the sodium crystal as containing a shared electron pair, moving around the body-centred atom in such a manner that each of the eight close neighbours contributes its 3*s* electron to that of the body-centred atom for one-eighth of the pseudo-covalent bond. The 3*s* electron taking part in this type of bonding is the outermost one, and Pauling refers to it as the 'valency' electron. This reference to valency is not that of the chemist, who uses the term to refer to combining power in forming ionic or covalent compounds.

The continual resonance of this type of bond implies that the outer or 'valency' electrons are being continually shared throughout the metallic lattice, for each of the atoms considered in the basic crystal of Figure 2.8 is also involved in resonance bonding with other atoms in adjacent cubes. This association prevails throughout the crystal lattice in three mutually perpendicular directions, and it becomes difficult to say exactly where any 3*s* electron from any one sodium atom is at any moment in time. For this reason, each atom must momentarily be bereft of its 3*s* electron, and become therefore a positively charged ion as in an ionic compound. This argument leads to the concept of metallic bonding as an array of positive ions surrounded by a 'free electron gas' of the 'valency' electrons continually moving through the lattice. The electron pairs form bonds between any two ions in turn, but are not permanently identified with any atom continuously. This type of bonding can thus be called 'unsaturated covalency bonding' when referring to the resonating shared electron pairs, and 'ionic bonding' with regard to the positive ions surrounded by an electron gas. The positive ions would normally repel each other, were it not for the electron gas holding them together by electrostatic attraction. It is as if the array of positive ions were packed as closely as possible, and yet were prevented from flying apart by the 'elastic bag' of electrons surrounding them and permeating the whole structure. The distance apart of the atoms will then depend on the type of packing (which depends also on their size) and the balance between attraction and repulsion forces, as with other types of bond.

This 'free electron' model applies to the elements at the beginning of each period, i.e. involving *s* electrons. The transition metals are more complex, in that the contribution of electrons to the Pauling 'valency' mechanism varies. This 'valency' is not necessarily an integer, since it varies according to the number of electrons to be paired off in spin and the type of crystal lattice adopted.

As the third group of metals is approached, those with one or two $3p$ electrons behave not unlike the alkali metals, although they are affected by the existence of s, d, or even f electrons in the energy levels closer to the nucleus. When we have two or more p electrons there is an increased tendency towards the covalent type of bond, since we are approaching the stable octet, and an obvious tendency to form such bonds by sharing electrons in a more permanent manner. This explains why some of the elements in this area are called 'semi-metals'. When p electrons predominate, either the covalent or ionic bond becomes the dominant factor. Hence the elements immediately preceding the rare gases (halogens for instance) are non-metals, i.e. they do not adopt the metallic bond.

The 'ideal' metallic bond, exemplified by the model described for sodium, requires the valency electrons to move continuously through the lattice. This inevitably involves a higher energy than for, say, a saturated covalent bond. The excess kinetic energy in the metallic bond is known as *Fermi* energy, and, being positive, constitutes a force of repulsion. The main force of attraction, mentioned above, is of the Coulomb (electrostatic) type between electrons and ions. Substituting in Mie's equation (equation (2.1)),

$$\phi(r) = -\frac{a}{r^m} + \frac{b}{r^n}$$

we can see that, for the ideal metallic bond, $m = 1$, as it is also for ionic compounds (Coulomb forces, equation (2.2)). The value of n is more difficult to obtain, but may be 2 for alkali metals, since Fermi energy varies theoretically as $1/r^2$.

For molecular lattices, where Van der Waals' energy is involved, $m = 6$ (equation (2.3)) and n appears to be equal to 7. For ionic compounds, $m = 1$ and $n = 9$.

2.8. *Implications of the type of bond on the structure of elements*

The inert gases, having stable electronic configurations, are normally monatomic, with very low boiling points. When liquefied, they associate into molecules having very weak binding, due entirely to Van der Waals' forces. In the solid state the binding increases slightly, but the temperature range over which these elements are either solid or liquid is very small indeed, which illustrates the weakness of the molecular bond. The boiling point increases with atomic number.

The halogen elements (fluorine, chlorine, bromine, and iodine) form molecules in the gaseous state by covalent bonding, i.e. they share one electron from each of two atoms. When liquefied, these diatomic molecules associate by Van der Waals' attraction, which, as usual, is comparatively weak. These substances are therefore highly volatile, i.e. have relatively low boiling points. The more complex the atom, however, the greater the polarization effects, and the Van der Waals' attraction increases. Fluorine, atomic number 9, boils at −187°C; chlorine ($Z = 17$) boils at −34·6°C; bromine ($Z = 35$) boils at +58·8°C. This illustrates the increased stability of the Van der Waals' bond with increase in atomic number. Iodine is a solid at room temperature.

Hydrogen, oxygen, and nitrogen have already been mentioned under covalent bonding. In the liquid and solid state, all three tend to form larger molecular associations by virtue of Van der Waals' attraction. In the case of hydrogen, this is obviously limited by the simplicity of the H_2 molecule, and hydrogen has a very low boiling point (−253°C). Oxygen and nitrogen are somewhat more complex, since they have directional bonds, with double and triple pairs of shared electrons respectively. Thus, the molecules attracted by Van der Waals' forces may branch out into longer molecules, rather like a string of beads. These complexities increase the Van der Waals' energy of attraction by the formation of dipole moments. (A reasonable analogy might be the magnetization of iron filings. Each filing becomes a little bar magnet, and is enmeshed with other filings in a weakly bound mass. If diatomic oxygen molecules act like the filings, they can enmesh also, but would be easily separated by rise in temperature.)

The boiling points of oxygen and nitrogen are thus higher than that of hydrogen, and oxygen has a higher boiling point than nitrogen (−183°C and −196°C respectively). The elements closely resembling nitrogen and oxygen in the next period (Table II) are phosphorus and sulphur. The increased atomic number gives rise to an increased Van der Waals' energy, and both these substances have melting and boiling points above room temperature, those for sulphur being higher than for phosphorus. Both form complex molecules in which the basic unit shows covalent bonding, and these units then associate by Van der Waals' attraction. Sulphur forms S_8 molecules making closed rings with a 'zigzag' periphery when in its rhombic form. Phosphorus and sulphur show allotropic modifications, i.e. each displays various crystal habits in the solid state, chiefly by virtue of varying Van der Waals' attraction.

Carbon, silicon, germanium, and tin form giant molecules, in which each atom is associated with four neighbours. In this case, the bonds are all covalent and are therefore stronger than those in the substances referred to above, which involve Van der Waals' forces to some degree in the solid state. The packing of atoms, however, is not the closest possible, so that the properties are very much affected by the type of crystal structure. This will be further discussed in the relevant chapters. Since these four elements are all in the same group of the periodic table (group IV) and all have four electrons short of the stable octet, it was recognized that the number of close neighbours must be $8 - N$, where N is the number of outer electrons, and also the group number of the periodic system. This follows, of course, from the formation of shared electron pairs of opposite spin. These elements also have allotropic modifications, so that some changes can take place in this rule. Carbon in the form of diamond obeys this rule, and so does 'grey' tin, i.e. the low-temperature form. Carbon gives rise to a host of possibilities in forming compounds by adopting single, double, or triple covalent bonds, and this forms the basis of organic chemistry. Silicon, nitrogen, and phosphorus do likewise, but exploration of the detailed chemistry and applications of these compounds is much more recent. Most scientists and technologists are aware, for instance, of the silicones and their application as lubricants and high boiling point oils. The nitrogen and phosphorus compounds are perhaps less useful and certainly less well known.

Arsenic, antimony, and bismuth are in the same group as nitrogen and phosphorus; selenium and tellurium are in the same group as oxygen and sulphur (see Table II). In the gaseous state, all five are stable diatomic molecules, with covalent bonding. All obey the $8 - N$ rule, so that arsenic, antimony, and bismuth, having five outer electrons, find three near neighbours in the solid state, and selenium and tellurium, where $N = 6$, have two. Neither of these types of structure can be closely packed, and the tendency is to form spiral chains or closed rings, as mentioned above for nitrogen, oxygen, phosphorus, and sulphur. These spirals or rings then are held together at least in part by Van der Waals' forces. Since the Van der Waals' force is greater than in the elements of the same groups in the early periods, there is an increase in melting point as atomic number increases. (The details of these structures are given in Chapter 3.)

The remaining metals, forming the bulk of the periodic system, pack more closely, and although the individual bonding energy is not

TABLE III

Melting and boiling points of the natural elements

Element		Melting point (°C)	Boiling point (°C)	Element		Melting point (°C)	Boiling point (°C)
Actinum	Ac	1,200†	3,300†	Neodymium	Nd	1,024	3,180†
Aluminium	Al	660·1	2,400†	Neon	Ne	−248·7	−246·1
Antimony	Sb	630·5	1,675	Nickel	Ni	1,455	2,910
Argon	A	−189·3	−185·9	Niobium	Nb	2,468	4,400
Arsenic	As	814	610‡	Nitrogen	N	−209·9	−195·8
Barium	Ba	710	1,700	Osmium	Os	3,045†	4,600†
Beryllium	Be	1,284	2,400	Oxygen	O	−218·8	−182·97
Bismuth	Bi	271·3	1,680	Palladium	Pd	1,552	2,980
Boron	B	2,050	2,550	Phosphorus	P	44·1	280
Bromine	Br	−7·2	61	(white)			
Cadmium	Cd	320·9	765	Platinum	Pt	1,769	4,170
Caesium	Cs	29·7	700	Polonium	Po	246	965
Calcium	Ca	850	1,350	Potassium	K	63·5	779
Carbon	C	3,700†	4,350†	Praseodymium	Pr	935	3,450†
(Graphite)				Radium	Ra	700	1,140
Cerium	Ce	804	3,600†	Radon	Rn	−71	−62
Chlorine	Cl	−101	−34·1	Rhenium	Re	3,180	5,600†
Chromium	Cr	1,850	2,680	Rhodium	Rh	1,960	3,780
Cobalt	Co	1,492	2,900	Rubidium	Rb	38·8	679
Copper	Cu	1,083	2,570	Ruthenium	Ru	2,450	4,200†
Dysprosium	Dy	1,500†	2,600†	Samarium	Sm	1,052	1,900†
Erbium	Er	1,500†	2,600†	Scandium	Sc	1,540	2,930†
Europium	Eu	826	1,700†	Selenium	Se	220	695
Fluorine	F	−219·6	−188·1	Silicon	Si	1,423	3,280
Gadolinium	Gd	1,350†	3,000†	Silver	Ag	960·8	2,193
Gallium	Ga	29·8	2,250	Sodium	Na	97·7	882
Germanium	Ge	937	2,870	Strontium	Sr	770	1,350
Gold	Au	1,063	2,950	Sulphur	S	119	444·6
Hafnium	Hf	2,220	3,100†	Tantalum	Ta	3,000†	6,000†
Helium	He	−271·4	−268·9	Technetium	Tc	2,200†	?
Holmium	Ho	1,500†	2,700†	Tellurium	Te	450	990
Hydrogen	H	−259·2	−252·8	Terbium	Tb	1,470†	2,800†
Indium	In	157	2,075	Thallium	Tl	304	1,457†
Iodine	I	113·5	183	Thorium	Th	1,750	3,850†
Iridium	Ir	2,443	4,620	Thulium	Tm	1,550†	2,400†
Iron	Fe	1,537	3,070	Tin (white)	Sn	231·9	2,750
Krypton	Kr	−157·8	−153†	Titanium	Ti	1,660	3,260
Lanthanum	La	920	4,515	Tungsten	W	3,380	5,400
Lead	Pb	327·3	1,740	Uranium	U	1,130	3,800
Lithium	Li	180	1,329	Vanadium	V	1,860	3,350†
Lutetium	Lu	1,700†	3,500†	Xenon	Xe	−111·8	−108·1
Magnesium	Mg	650	1,105	Ytterbium	Yb	824	1,800†
Manganese	Mn	1,244	2,095	Yttrium	Y	1,550†	3,030†
Mercury	Hg	−38·9	356·6	Zinc	Zn	419·5	907
Molybdenum	Mo	2,625†	5,690†	Zirconium	Zr	1,860	4,750†

† approximate ‡ sublimes

always so very high in the metallic bond (compared with ionic and covalent compounds), the close packing ensures a relatively high strength. The melting and boiling points usually increase with atomic number in any group and are highest in the transition metals. Tungsten, atomic number 74, has the highest melting point. The lanthanide and actinide series of elements (having outer *f* electrons) have some unusual properties. Table III gives a list of melting and boiling points.

The properties of elements, in particular, and of compounds, are affected by the crystal lattice, and it is essential to discuss the characteristics of these lattice types before dealing with the properties in detail. This will be done in Chapter 3. There is, however, one property of great importance that is fundamentally affected by the type of bond, and this is discussed in the final section of this chapter.

2.9. *Implications of the type of bond on conductivity*

Discussion of the metallic bond (2.7) showed that in the alkali metals, such as sodium, the hybridization of *s* and *p* energy levels, and the contribution of only one valency electron from each sodium atom, inevitably leaves unoccupied states of energy in the sodium atoms of the lattice. This outer band of electron states is frequently called the 'conduction' band, for it is here that electrons can change energy when there are unoccupied states and can therefore pass on electric current. An applied electric field, in these circumstances, causes an immediate resultant flow of electrons, which is what is meant by an electric current. One definition of a metallic structure is therefore the possession of an incompletely filled conduction band of electrons. In contrast, diamond, having four electron pairs completely filling the conduction band, has no unoccupied states, and consequently the diamond structure is an insulator. The perfection of the crystal lattice is highly significant in the matter of conductivity, even with the metallic lattice. It is easy to see that if the electrons must have perfect freedom of movement through the lattice for high conductivity, any obstruction to their movement will interfere with the conductivity. The obstructions may be impurity atoms, which cannot form part of the bond without some disturbing effect on the electrons, even if the impurity is another metal. The purer the metal, therefore, the higher the electrical conductivity. The obstructions may also be physical barriers, such as increased distance between the atoms. Thermal energy causes the atoms to vibrate (see Chapter 5) and the amplitude of vibration increases with rise in temperature. The movement of electrons is now through a slightly different lattice, and consequently the passage is rendered more difficult by rise in temperature and conductivity decreases. This explains the normal increase in resistance of a metal when heated, giving rise to a positive temperature coefficient of resistance. The more pure the metal, the less this temperature coefficient and the lower the residual resistance at low temperature.

The overlap of energy levels illustrated for magnesium in Figure 2.1 can serve as a guide to the conductivity of a material. It is necessary to know whether the overlap takes place, and the concentration of electrons in that band. A full band will not allow conductivity, a partially occupied one will. There are more subtle arguments involved, however, in explaining the behaviour of the semiconductors, and a more detailed look at the conduction band will do this.

The distribution of electrons can be studied in more detail by

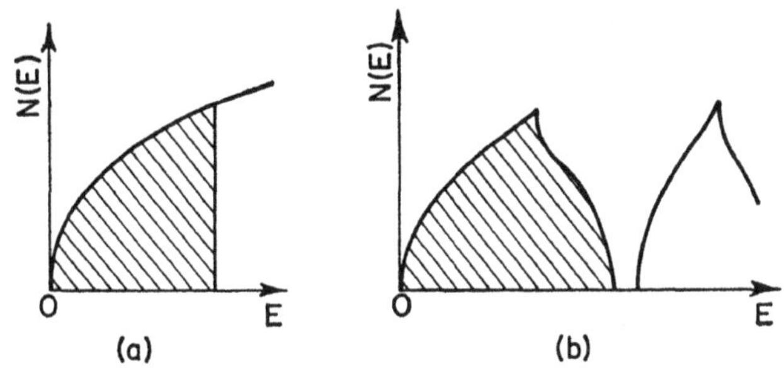

FIG. 2.9. Density of states (schematic). (*a*) Sodium. (*b*) Carbon (diamond).

constructing the density-of-states curve, often known as the $N(E)$ curve. A quantity $N(E)$ is defined such that $N(E)\mathrm{d}E$ is the number of energy states per unit volume with energy ranging from E to $E + \mathrm{d}E$. When $N(E)$ is plotted against E, the shape of the curve indicates the available states, and the actual occupation of these by electrons can be shaded in (see Figure 2.9a). (For convenience, the occupation of states is referred to absolute zero, which simplifies the mathematical reasoning and eliminates temperature effects for the moment.) The example given shows one in which the number of states of given energy increases with increase in energy. The shaded area is less than the available area, so that there are unoccupied states in this case. The crystal is therefore an electrical conductor. This type of curve is typical of metals that approach the free electron model, such as sodium. With some other atoms, the crystal lattice modifies the distribution of electrons, and the density-of-states curve has more the shape shown in Figure 2.9b. This merely indicates a boundary occurring at a particular energy level, a gap in which no electrons can exist, and a further band of electron energy states following at a higher energy level. In the example given, the shaded area occupies the whole of the area under the curve, so that the first available states are all filled. There is no possibility of moving into unfilled states except by crossing a forbidden energy gap. The substance is normally such an insulator. $N(E)$ curves (Figure 2.9b) merely illustrate the density of states that have been argued for carbon as diamond, or any other covalent bond that achieves the stable octet, and explain the lack of electrical conductivity.

It is obvious, however, that the size of the forbidden energy gap in Figure 2.9b is of great importance. If it is small, as in Figure 2.10a, then, since these curves refer to absolute zero, the substance will be an insulator at that temperature. When temperature increases, however, there is the possibility that the increased vibrational energy may cause an electron to cross this small energy gap and enter the unoccupied states of the next band. Further increase in temperature will allow more electrons to cross the gap, until all states in the second band are occupied. Furthermore, the states in the first band are now no longer fully occupied. The net result is an increase in conductivity as temperature rises, a negative temperature coefficient of resistivity. Such substances are known as *intrinsic semiconductors*, because they are insulators below a given temperature and conductors with a negative temperature coefficient above it. Silicon and germanium are examples of this type of semiconductor.

There is a second type of semiconductor, however, that would be an insulator in the very pure state, but whose characteristics can be modified by impurities. These substances are therefore known as *impurity semiconductors*. Figure 2.10b shows an $N(E)$ curve for such a substance, in which the forbidden energy gap is quite wide but the presence of impurity atoms has contributed an $N(E)$ curve of energy states lying between those of the basic material. Figure 2.10b again represents conditions at absolute zero. As the temperature is raised, electrons cannot jump the gap between the permitted states of the basic insulator, but electrons from the impurity atom states may be able to make the jump into the unoccupied band. Conductivity will then result. In time, all the available electrons from the impurity levels have entered the conduction band. In this case, further rise in temperature may cause a drop in conductivity, from the interference in the periodicity of the electron field. Thus, once conductivity commences at the appropriate temperature, there is initially a negative temperature coefficient of resistivity (rising conductivity) and then a transition to a positive temperature coefficient as for normal metals, when conductivity falls. This type of semiconductor is known as an *n-type*, or *donor*, because the impurity atoms supply the electrons, and these enter an unfilled band of states and act as normal free electrons (n for negative charge).

In another type of semiconductor, the impurity atom energy levels are *acceptors* rather than donors, and the effect of temperature is to excite electrons from the filled band across the gap into the impurity levels. Unoccupied states then remain in the previously filled band, and once more the energy of the electrons still in this band can be redistributed, thus restoring conductivity. The acceptor impurity gives rise to a somewhat different situation, as if 'positive holes' were created for the electrons. This substance is therefore called

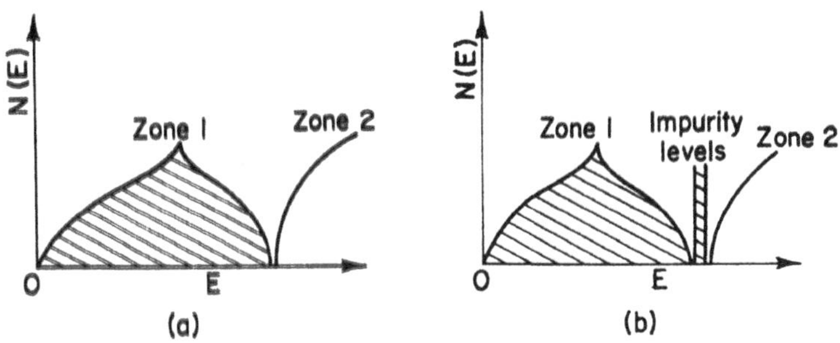

FIG. 2.10. Density of states. (*a*) Intrinsic semiconductor. (*b*) Impurity semiconductor.

a *p-type semiconductor* or acceptor. (p denotes positive holes provided by the impurity.) Once the impurity atoms are saturated with electrons they can accept no more; the increasing amplitude of vibration of the thermally excited electrons in the conduction band then begins to affect the conductivity as with a normal metal, and there is a positive temperature coefficient of resistivity. Impurity semiconductors therefore show the peculiarity of negative temperature coefficient initially, and positive later.

As referred to above, silicon and germanium are intrinsic semiconductors, being crystals of the diamond-type structure with four valency electrons per atom. When phosphorus or arsenic are added, these atoms contain five valency electrons per atom, and electrons from these can enter the conduction band, thus creating an n-type semiconductor. Boron or gallium, with three valency electrons per atom, cannot supply electrons to the conduction band in silicon or germanium, but boron and gallium can accept them from the filled band below. Boron and gallium, therefore, can give rise to p-type semiconduction in silicon or germanium.

By ringing the changes between silicon or germanium alone, acting as intrinsic semiconductors, the addition of phosphorus or arsenic as n-type impurities, or the addition of boron or gallium as p-type impurities, the variation of electrical conductivity with temperature can be adjusted over a very wide range, and combinations of these types can be devised to act as electronic devices with powers of rectification or amplification. These substances therefore are the basis of many electronic applications.

3

Constitution

3.1. *Crystallographic principles*

In the light of the distribution of electrons discussed in Chapter 1 and the modifications for the valence state mentioned in Chapter 2, the atom is obviously not a hard sphere. However, the restrictions of wave mechanics and the quantum theory indicate that a roughly spherical energy distribution for the electrons can be envisaged for the purpose of explaining the packing of atoms. Representation of atoms by spheres is therefore a convenience rather than a fact,

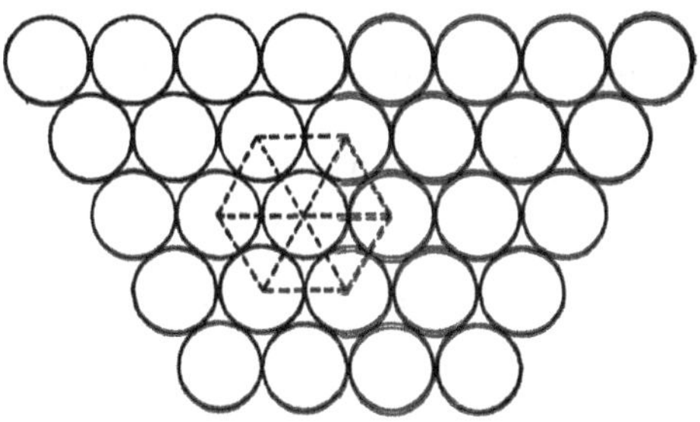

FIG. 3.1. A closely packed plane of spheres (hexagonal symmetry).

though it does assist in some quantitative discussion of the lattice characteristics.

The closest possible packing of spheres in a plane can be illustrated quite simply. Plate IV is a photograph of such a plane array, and Figure 3.1 indicates the basic unit from which the whole plane is built, one sphere surrounded by six others, each touching the central sphere and two others in its own outer ring. This type of close-packing has a hexagonal symmetry, as can be seen from Figure 3.1. If the spheres are slightly rearranged as if occupying squares on a chess board (Figure 3.2), then each sphere has four close neighbours only, and the packing cannot be of the closest type, although it is still close. A crystal lattice is three-dimensional, so that the planes shown in Figures 3.1 and 3.2 must be extended in a direction perpendicular to those planes to reproduce the complete lattice. Taking the square packing of planes in Figure 3.2, planes of identical packing can be added above and below this basic one, such that every sphere is in the same vertical axis as its neighbours in the planes above and below. This creates a simple cubic lattice, in which each sphere is equidistant from six near neighbours, four in its own plane and one each in planes above and below. Figure 2.3 illustrates this, with reference to the NaCl-type structure. The coordination number, i.e.

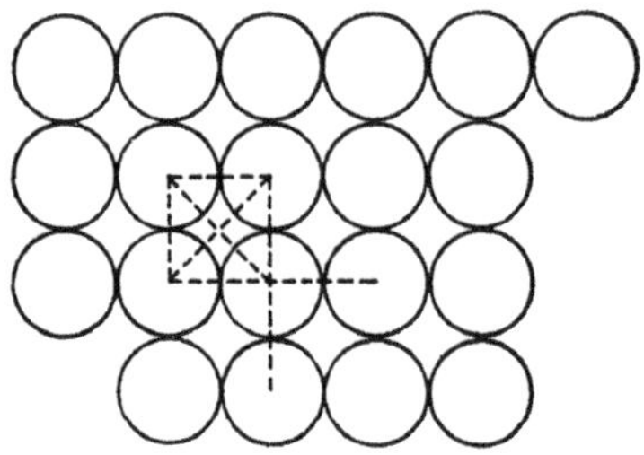

FIG. 3.2. A square-packed plane of spheres.

the number of nearest neighbours, is obviously 6, and this cannot be the closest form of packing. It is still closer, however, than the diamond lattice of Figure 2.6, which is tetrahedral in form, and has a coordination number of only 4. Just as the square is the unit of the plane in Figure 3.2, so the simple cube, with a sphere at each corner of the cube, must be the *unit cell* of the structure in three dimensions. In many cases it is convenient to avoid drawing the complete crystal lattice extended over many unit cells, and merely to draw the unit cell itself. Atoms in a simple cubic cell occupy 52% of the available space, those in the diamond structure 34%.

Figure 3.1, showing the closest possible packing in a plane, may now be built up into three dimensions. The second layer or plane of spheres can be placed on the first so as to occupy each space between three spheres in the lower plane, as in Figure 3.3. This second plane is identical to the first in close-packing, but has moved at an angle to the vertical axes through the spheres of the first layer so as to pack closely into the available space. It is in placing the third layer

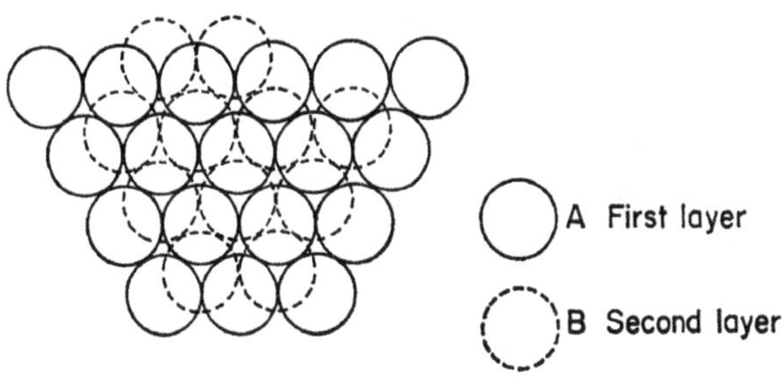

FIG. 3.3. Two layers of close-packed planes.

that differences can now arise. The third layer may be placed immediately above the first, i.e. with the third-layer spheres on the same vertical axes as the first. This gives a packing best described as A B A B A B . . ., since the third and first layers are identical in three mutually perpendicular directions, but layer B is identical only within the original plane, the third axis being displaced with respect to a vertical axis at an angle less than 90° to the upper and lower layers (see Figure 3.4a).

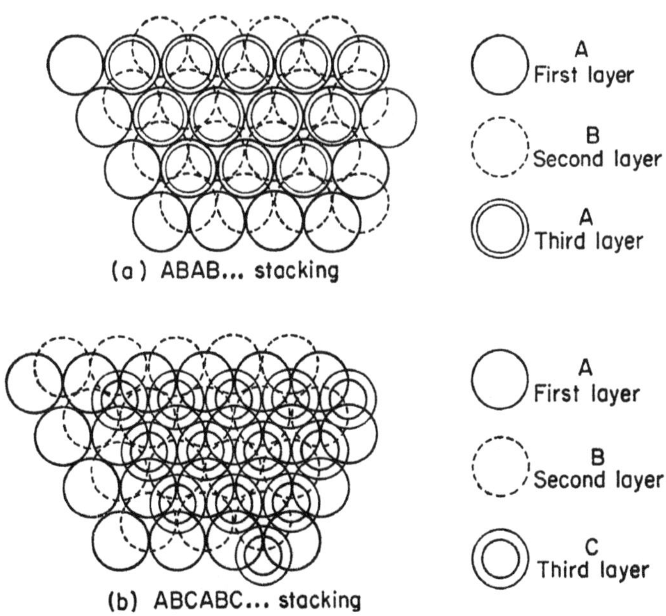

FIG. 3.4. Three layers of close-packed planes. (*a*) A B A B . . . stacking. (*b*) A B C A B C . . . stacking.

The second method of packing the third layer is not to place it directly over the spheres of the first layer, but in the spaces between the spheres of the second layer and again displaced with respect to a vertical axis. The fourth layer, following this one, would similarly be placed over the spaces in the third layer, but these are now directly above the spheres of the first layer, and so the packing repeats again as a pattern of A B C A B C . . . (see Figure 3.4b). The difference between these two structures is illustrated by the rather more 'open' diagrams of Figure 3.5. Both these structures represent

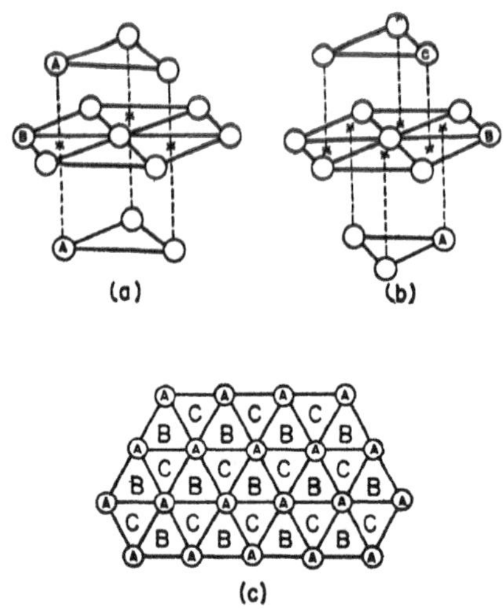

FIG. 3.5. The difference in packing of the layers in (*a*) close-packed hexagonal; (*b*) face-centred cubic (after Zwikker); (*c*) face-centred cubic (after Hume-Rothery).

the closest possible packing of spheres in three dimensions. If a rigid container is filled with spheres in either of these closest-packed arrangements, the spheres occupy 74% of the available space. From Figure 3.5 it can be seen also that each sphere has twelve close neighbours, a coordination number of 12.

The examples of packing mentioned so far range from coordination number 4, for the tetrahedral diamond type, to 6 for the simple cube and 12 for the two closest-packed types. There is one other important mode of close-packing, a development of the simple cube. By adding an identical second layer to the plane of Figure 3.2, but allowing the second-layer spheres to occupy the space between four spheres of the lower square lattice, the A B arrangement becomes more closely packed. A third layer may now be added, with the spheres vertically above those in the first layer, to give A B A B . . . again, but this time in the square lattice, not the close-packed hexagonal one (Figure 3.6). The resultant lattice is that of Figure 2.8, and is called the *body-centred cubic* structure. The coordination number is 8, as explained for sodium in Chapter 2, and hence this structure falls between the simple cubic (6) and the closest possible packing (12).

The two types of closest packing are referred to as *face-centred cubic* (Figure 3.4b) and *close-packed hexagonal* (Figure 3.4a). The

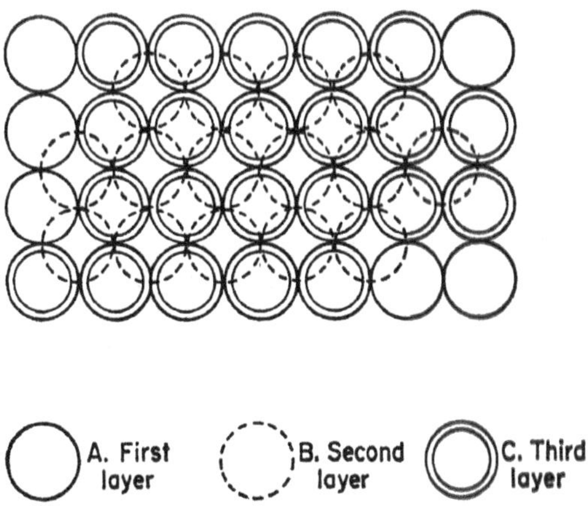

FIG. 3.6. Three layers of square-packed planes A B A B . . . stacking (b.c.c.).

face-centred cubic unit cell contains one sphere at each corner of a cube and one at the centre of each face, hence the name. It is not easy to see the derivation of the unit cell from the close-packed layers shown in Figure 3.4b, but it can best be appreciated in the reverse direction. Plate V shows a photograph of a face-centred cubic cell, from which a sphere at one corner has been removed so that the spheres lying in the plane are revealed. It will be seen that these are close-packed and in fact show the build-up of the A B C layers that constitute the structure. Figure 3.5a illustrates the features of the close-packed hexagonal structure reasonably clearly, since the cell can be envisaged as a hexagonal ring of spheres with one in the centre (the closest possible packing in one plane), an intermediate layer of three atoms in three of the available close-packed spaces, and a third layer, identical with the first, to complete the packing.

The unit cells described account for the majority of structures in the elements other than those of homopolar nature or involving Van der Waals' bonding. Many compounds, especially the ionic compounds, crystallize in more complex forms, as in the case of the natural minerals. To study the basic principles of materials, it is not essential to explore fully the crystallography of all these types, but the most common lattices may be examined in more detail in order to appreciate some of the wide variations in properties and the *limitations* of materials that automatically stem from both their natural or man-made forms. The closeness of packing of atoms in a crystal lattice affects thermal and mechanical properties, and for this reason is particularly important in the metals, not only for influences on possible applications but also requirements for fabrication into shape.

3.2. *The unit cell*

As described above, the packing of atoms involves a three-dimensional array resulting from the repetition of a unit cell. The consideration of unit cell types serves to classify structures into groups, and to correlate similar patterns of behaviour and intrinsic properties in these groups.

The crystallographic axes to which the unit cell is referred are designated by x, y, and z, and are related by the angles α, β, γ between them (Figure 3.7). The point P in the lattice of unit cell of sides a, b, c can be described by the vector sum $OP = u_a + v_b + w_c$ (where u, v, w are the steps along the x, y, z axes). Any direction in such a lattice

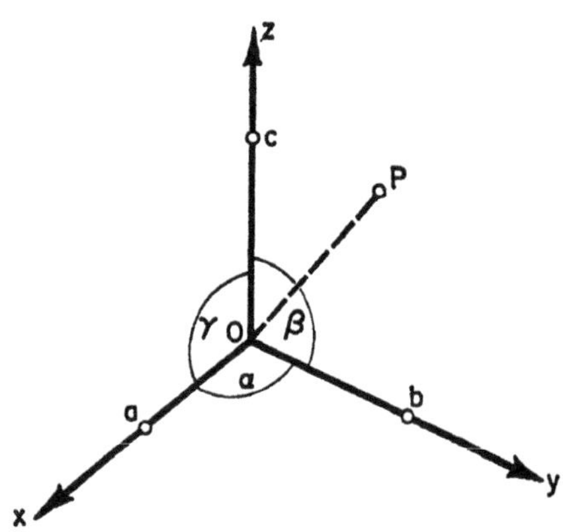

FIG. 3.7. Crystallographic axes.

can thus be defined by the general expression [*uvw*]. The square brackets are used to indicate a direction, and *u*, *v*, and *w* are reduced to the smallest possible integers for convenience.

When a plane is to be described, the notation used is that of *Miller indices*. The plane is extended to cut the *x*, *y*, and *z* axes, and the intercepts in steps of the unit cell dimensions *a*, *b*, *c* are written down. The reciprocals of these intercepts are next obtained, and

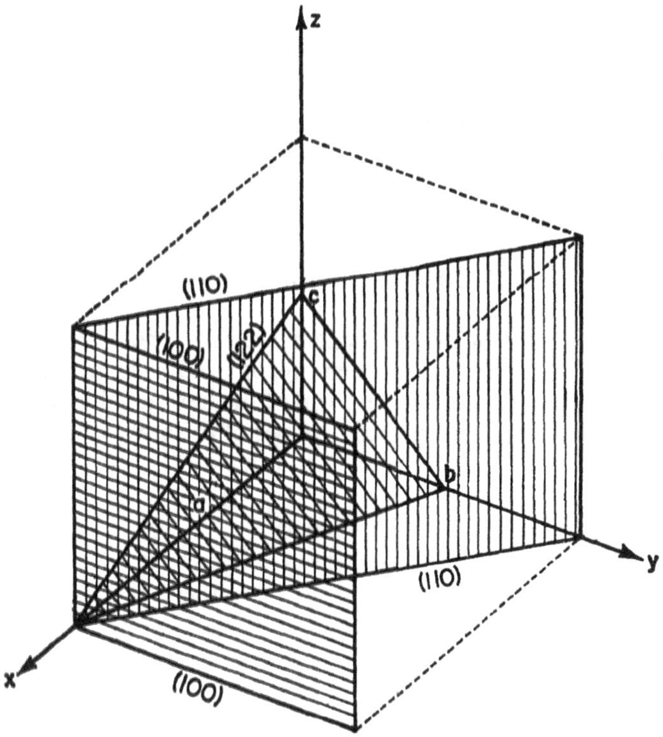

FIG. 3.8. Miller indices of some planes in the cubic system.

fractions cleared to give the smallest integers again. Figure 3.8 shows three such planes in the cubic system where $a = b = c$, (i) parallel to the *y* axis and the *z* axis, (ii) parallel only to the *z* axis, and (iii) intercepting all three axes. In turn, the intercepts on plane (i) are 2 on the *x* axis, infinity on the *y* and *z* axes.

The intercepts are written	2;	∞;	∞
Reciprocals taken:	$\frac{1}{2}$;	0;	0
Remove fractions:	1	0	0

This plane is therefore known as the (100) plane, the parentheses indicating a plane.

Intercepts in plane (ii) are	2;	2;	∞
Reciprocals taken:	$\frac{1}{2}$;	$\frac{1}{2}$;	0
Remove fractions:	1	1	0

Plane (ii) is therefore a (110) plane.

For plane (iii)

Intercepts	2;	1;	1
Reciprocals taken:	$\frac{1}{2}$;	1;	1
Remove fractions:	1	2	2

Plane (iii) is therefore a (122) plane.

Typical planes, with their Miller indices, are illustrated in the solid models of Plate VI. In the hexagonal crystal system it is more convenient to use four axes than three, to emphasize the hexagonal symmetry. The z axis (side c of the unit cell) is retained, but three axes at 120° to each other replace the x and y axes, and so define the side a of the unit cell.

The basic types of crystal system are few. There are six forms, although in the past others were recognized. The rhombohedral system, for instance, is a special form of the hexagonal, and was once classified separately. The essential values of the unit cell sides a, b, c, and the angles α, β, γ, are shown in Table IV.

TABLE IV

Crystal systems

System	Unit cell	
Isometric (cubic)	$a = b = c$	$\alpha = \beta = \gamma = 90°$
Hexagonal	$a = b \neq c$	$\alpha = \beta = 90°\quad \gamma = 120°$
Tetragonal	$a = b \neq c$	$\alpha = \beta = \gamma = 90°$
Orthorhombic	$a \neq b \neq c$	$\alpha = \beta = \gamma = 90°$
Monoclinic	$a \neq b \neq c$	$\alpha = \beta = 90° \neq \gamma$
Triclinic	$a \neq b \neq c$	$\alpha \neq \beta \neq \gamma$

3.3. *The face-centred cubic lattice (f.c.c.)*

Plate V, showing the close-packed planes of this system, reveals the octahedral, or (111), planes, since there are equal intercepts on all three axes a, b, c. The direction of closest-packed rows of atoms is [110]. There are four sets of closely packed (111) planes in the f.c.c. structure, and twelve sets of close-packed directions. Since plastic deformation of crystals frequently occurs by slipping of close-packed planes over one another, and in the close-packed directions, the face-centred cubic structure might be expected to show considerable plasticity. This is indeed the case, and examples of this structure include the metals aluminium, copper, silver, and gold, all of which are ductile.

The coordination number of the f.c.c. lattice is 12, and if a is the length of the unit cell side ($a = b = c$) then the nearest neighbours are situated at a distance of $\frac{1}{2}a\sqrt{2} = 0{\cdot}707a$. There are also near neighbours (six to each atom) at a distance a. Voids therefore exist

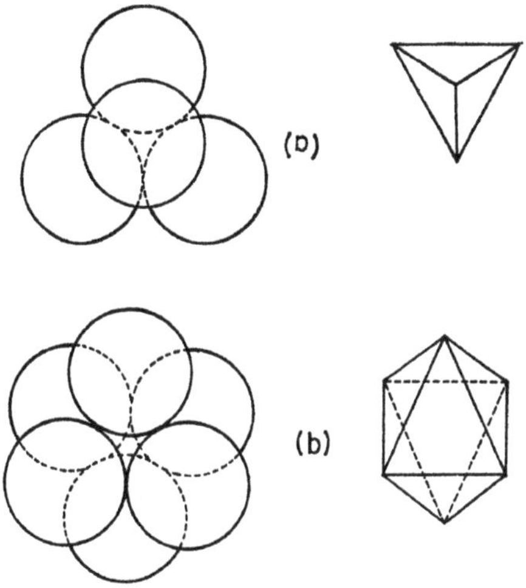

FIG. 3.9. (*a*) Tetrahedral voids. (*b*) Octahedral voids.

in the structure, and other smaller atoms may fit into these spaces. This is particularly important in inorganic crystals, as well as in certain metals. If we refer to the close-packed layers in Figures 3.4 and 3.5, a tetrahedral void occurs when one atom is situated over the triangular space in the lower layer between three atoms. Figure 3.9a shows that four atoms exist around this void, hence the tetrahedral shape. The arrangement in Figure 3.9b is one in which a triangular space in the first layer connects with a similar triangular space in the upper layer, so that six atoms are involved, arranged on the corners of an octahedron. The void between these atoms must also be octahedral. Octahedral holes can just contain a sphere of radius $0{\cdot}414R$, where R is the radius of the primary spheres of the lattice, and tetrahedral holes can just contain a sphere of radius $0{\cdot}225R$. The f.c.c. lattice obviously has the largest holes at the centre of the cube, and at the centres of each cube edge (Figure 3.4). The tetrahedral holes in a f.c.c. cell are situated towards each corner of the unit cell cube, between a corner atom and the three face-centred atoms adjacent to the corner. Since there are eight corners, there are eight tetrahedral holes (Figure 3.10). (The majority of elements entering interstices in

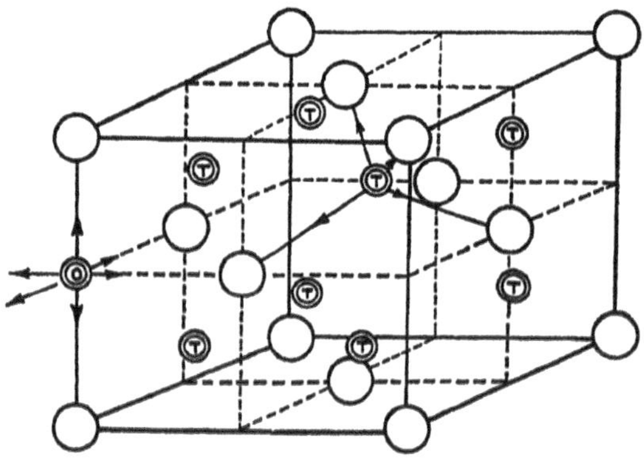

FIG. 3.10. Voids in face-centred cubic structure.

crystals are larger in diameter than the holes described, and so distort the lattice; see page 185.)

The face-centred cubic structure is adopted by many metals and, as already mentioned, these show considerable plasticity, owing to the close-packing in the planes on which deformation takes place and the number of close-packed directions. The *lattice parameter*, or unit cell side, varies according to the size of the atom concerned and its particular electronic structure. Lead, for example, has a f.c.c. cell of side $a = 4{\cdot}94$ kx, whereas copper, also f.c.c., has $a = 3{\cdot}61$ kx. kx are units of crystal size based on the crystal of calcite. Multiply kx units by 1·00202 to give ångström units (Å). 1 ångström $= 10^{-8}$ cm. As mentioned above, close-packed spheres occupy 74% of the available space in such a lattice.

3.4. *The close-packed hexagonal lattice (c.p.h.)*

The c.p.h. structure has been dealt with in discussing the possibilities of close-packing, and is illustrated by Figures 3.4a and 3.5. This lattice is somewhat simpler to envisage than the f.c.c. which it so closely resembles, since the packing of A B A B . . . layers and the hexagonal symmetry of each layer can readily be demonstrated with everyday objects. The close-packed layers A and B in this stacking sequence are identical, but they are the only close-packed planes, as compared with the f.c.c. octahedral planes. Thus, since close-packed planes are likely to be the site of atomic movements in deformation, the c.p.h. structures show deformation mainly in the *basal plane*, i.e. the close-packed A and B layers. The close-packed directions will again lie in these planes. The Miller–Bravais notation, utilizing three *a* axes of equal length at 120° to each other in the hexagonal close-packed plane, and a fourth *c* axis at right-angles to the other three, describes the basal plane as (0001), since the plane is parallel to each of the three *a* axes and intercepts only the *c* axis. In this notation, the first three numbers relate to the *a* axis and the fourth to the *c* axis. Here the sum of the rationalized intercepts on the *a* axes (Miller indices) must be zero, a special feature of this system. One of the three close-packed directions in the basal plane is therefore $[11\bar{2}0]$, the others being $[\bar{2}110]$ and $[1\bar{2}10]$. A less closely packed plane in the c.p.h. lattice is that joining two A layers in a vertical plane, a 'prismatic' face of the basic cell. There are six such planes in the cell, all having the last index in the Miller notation equal to 0, since they are parallel to the hexagonal axis. Each of the six is also parallel to one of the three *a* axes, and has an intercept on each of the other two, one positive and one negative. Figure 3.11 illustrates this on both cell and plane diagrams. The planes are therefore set out in three sets of parallel faces, viz. $(01\bar{1}0)$ and $(0\bar{1}10)$; $(10\bar{1}0)$ and $(\bar{1}010)$; and $(1\bar{1}00)$ and $(\bar{1}100)$. The fact that these planes are much less closely packed is only part of the story. It will be obvious

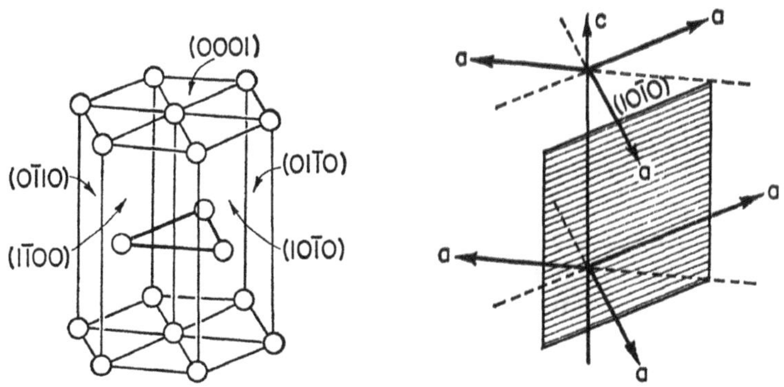

FIG. 3.11. Close-packed basal and prismatic planes, hexagonal system.

from Figures 3.11, 3.4a, and 3.5 that the atoms in the basal-plane layers A are much closer together than the basal-plane A layers themselves. Hence the prismatic face is rectangular in outline. The c.p.h. structure might be expected to be less plastic than the f.c.c., and this is indeed so. Furthermore, it will exhibit 'anisotropy', i.e. different behaviour along different crystallographic axes, because of the 'stratification' of the closely packed basal planes and the more open packing in the prismatic ones.

In the close-packed hexagonal lattice each atom in the basal plane has six neighbours at a distance a, and neighbours in the planes above and below at a distance $\sqrt{(\frac{1}{3}a^2 + \frac{1}{4}c^2)}$, where c is the hexagonal axis separation of the individual A layers in the basal plane. In the case when $c/a = \frac{1}{3}\sqrt{8} = 1{\cdot}6333$, separation of nearest atoms in

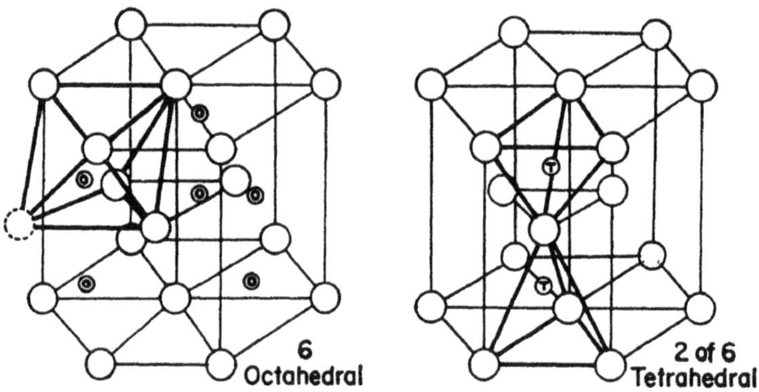

FIG. 3.12. Voids in the close-packed hexagonal system.

planes above and below is equal to a, and all nearest neighbours (twelve) are at the same distance. Six of these are in the same plane as any individual atom, and three are in planes above and below. The next nearest neighbours are in the planes above and below but one atom removed from the nearest neighbours in those planes (see Figure 3.11). The distance between these second close neighbours is $a\sqrt{2} = 1{\cdot}414a$. The third closest are the atoms in prismatic planes, and these are distance c apart ($1{\cdot}633a$).

As explained previously, two types of void exist in the close-packed structures. In the c.p.h. structure the octahedral holes are situated between and below two atoms in the basal plane (along an a axis) and between and above the two nearest atoms in the plane below (B layer). There are three such positions above layer B and three below in each unit (Figure 3.12). Similarly, the tetrahedral holes are directly above an atom in the B layer, and equidistant from three atoms in the basal plane, i.e. not along an a axis. (There are three such positions above and three below each B layer: see Figure 3.12.) For a close-packed structure, the sizes of the voids are always in the same ratio to the atomic radius (regarded as a sphere) whether f.c.c. or c.p.h.

It is not essential that all lattices of hexagonal symmetry should be close-packed. As already emphasized, the atoms are not hard spheres and the symmetry of the electron cloud is frequently more complex. Interaction of the outer energy levels produces repulsive forces which naturally vary according to the size of the atom and its electronic characteristics. The interatomic distance may therefore exceed the value expected in the close-packed lattice. This becomes most evident in the hexagonal structures, which are not always close-packed. If the hexagonal axis c is greater than 1·633, then the structure is still hexagonal, but not the closest-packed. The properties will be altered, for not all the near neighbours of atoms in the basal plane will be equally near, and the spacing of the prismatic planes will be increased in the c direction. This actually occurs with some of the metals of hexagonal structure. Some, indeed, have a c/a ratio less than 1·633; beryllium is a good example, with $c/a = 1{\cdot}63$ in its high-temperature form. Zinc and cadmium have high values, 1·86 and 1·89 respectively, and so are extended in the c direction. In non-closed-packed hexagonal lattices, the space occupied by the atoms is less than the 74% of that in close-packing.

The relationship between f.c.c. and c.p.h. structures is so close that transition from one to the other takes place in certain metals when thermal effects may allow some rearrangement. Cobalt, for example, exists as a c.p.h. lattice at low temperatures, and as a f.c.c. one at high temperatures. Strontium has a f.c.c. lattice at low temperatures, a c.p.h. lattice at higher temperature, and changes again nearer to its melting point. The similarity of the two structures also allows a partial transition over a restricted number of planes. If the A layer of the c.p.h. lattice is moved through a small distance, it can become a C layer as in the f.c.c. system. This occurs only over a small distance, and requires particular conditions to produce such movements; such conditions will be referred to in Chapter 4. When such a change in sequence occurs, it is referred to as a 'stacking fault'. Analogous faults can occur in the basic f.c.c. structure. The energy of these stacking faults varies from one metal to another (of those that crystallize in the close-packed systems) and so the probability of their occurrence may affect variations in properties.

3.5. *The body-centred cubic lattice (b.c.c.)*

This type of lattice has already been mentioned, particularly in the discussion of the free electron model of sodium (see pages 38–40). The structure is obtained by an A B A B . . . type of packing on a square lattice (Figure 3.6). The most closely packed planes are obviously those that include the body-centred atom and the opposite edges of the basic cube. These are the (110) or dodecahedral planes, and there are six of these per unit cell. The closest-packed direction is that of [111], and there are four such directions. Since these closest-packed planes are still more open than those of the f.c.c. lattice, and there are fewer close-packed directions, the b.c.c. structure cannot show the same plasticity as the f.c.c. Thus the b.c.c. metals tend to be more brittle than the f.c.c. ones. Atoms in a b.c.c. structure occupy 68% of the available space. The (112) and (123) planes are also sufficiently close-packed to be involved in deformation by slip. The b.c.c structure has a coordination number of 8, and each atom is surrounded by eight others at a distance of $\frac{1}{2}a\sqrt{3} = 0{\cdot}866a$, where a is the side of the unit cell. There are also six near neighbours at distance a. The near neighbours are thus only 15% farther away than the closest neighbours, which is much closer than the equivalent nearest and next-nearest atoms in the closely packed structures, although they, of course, have twelve nearest neighbours compared with the eight of the b.c.c. system. The b.c.c. structure can also be regarded as two interpenetrating simple cubic lattices, the body-

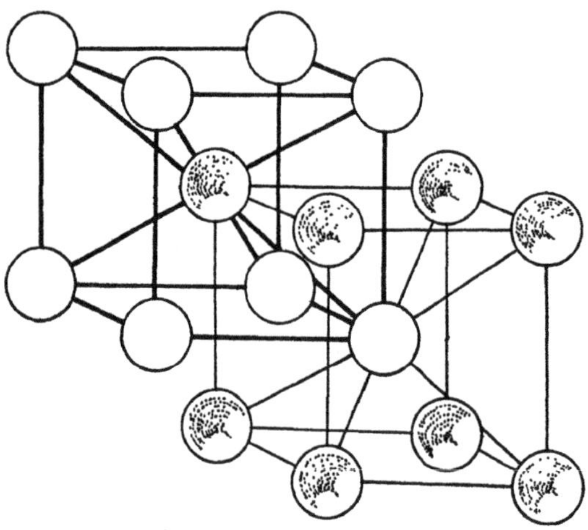

FIG. 3.13. Interpenetration of cubes in the body-centred cube structure.

centred atoms of the main lattice being in simple cube relationship with each other (Figure 3.13). This is important when considering the crystallization of some inorganic compounds in the b.c.c. system, and the 'ordering' of alloys (Chapter 8).

If two adjacent unit cells of a b.c.c. structure are considered, the distribution of voids in the structure becomes more apparent. Octahedral holes occur at the centre of cube faces, and at the centre of cube edges. They are nevertheless irregularly shaped voids, since

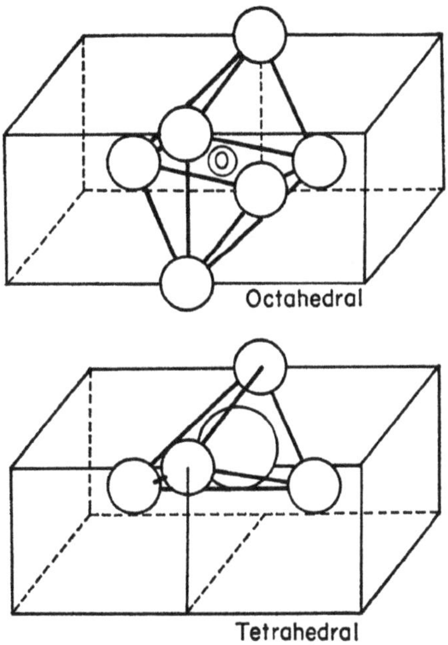

FIG. 3.14. Voids in the body-centred cubic structure (after Hume-Rothery).

two atoms are at a distance $\frac{1}{2}a$ and four at $a/2\sqrt{2} = 0{\cdot}71a$. A sphere of radius $0{\cdot}154R$ can fit into these spaces. The tetrahedral holes occur in the space between the two body-centred atoms and the two closest cube corners at the face common to the two cells (Figure 3.14). There are thus four such holes in each cube face. The sides of the tetrahedral hole are two of length a and four of length $a/2\sqrt{3} = 0{\cdot}866a$. The tetrahedral are therefore larger than the octahedral holes, and can contain a sphere of radius $0{\cdot}291R$. The tetrahedral holes in b.c.c. structures are slightly larger than the corresponding holes in close-packed structures, but much smaller than octahedral holes in the latter. The opportunity for elements with small radius atoms to occupy the holes in a b.c.c. lattice is therefore reduced, and this fact can be of immense importance. It is one contributory factor to the hardening of steel, since carbon atoms can more readily fit into the f.c.c. lattice of iron that exists at higher temperatures than into the b.c.c. structure that exists at room temperature. When carbon steels are heated to temperatures above 760°C, the carbon atoms enter the octahedral holes in the f.c.c. (γ) iron. On quenching from this temperature into water, they cannot remain in the equivalent positions as the crystal structure tries to change back into b.c.c. (α) iron. Quenching allows no time for equilibrium to be restored and so gives rise to a transition structure that is under severe lattice strain.

3.6. *Twin crystals*

Atoms in a crystal may move in such a way as to create a new crystal related to the original by symmetry about a plane. In the metals, the two portions of the crystal are often mirror-images of one another across this plane, known as a *twin plane*. Figure 3.15a shows the two-dimensional lattice of a twin crystal, in which a line represents the plane of the three-dimensional model. Each point on the lattice moves through a distance proportional to its distance from the twinning plane. A shearing mechanism can produce such a change, and Plate VII shows how this can take place by rotation of part of a cubic crystal.

Twinning can therefore provide a plastic deformation mechanism quite different from that of slip. In slip, a restricted number of planes are involved, and a restricted number of directions, so that whole areas of the crystal are affected. In twinning, each plane of atoms moves through the same distance relative to the neighbouring planes. Face-centred cubic structures twin on the close-packed (111) planes (Plate V), and the twins are extremely stable, remaining in the structure even after prolonged annealing. This explains their appearance in the so-called 'stainless' steels, which have the structure of gamma iron, face-centred cubic. Body-centred cubic structures also show twinning, on the (112) plane. This plane is not close-packed, and hence not all the atoms will be in the proper plane after twinning. Internal strain will therefore exist across the mismatched lattice (Figure 3.15b). High-temperature annealing will remove this strain. In alpha iron (b.c.c.) impact loading at room temperature can produce shear twins, called *Neumann bands* (which are found, incidentally, in metallic meteorites).

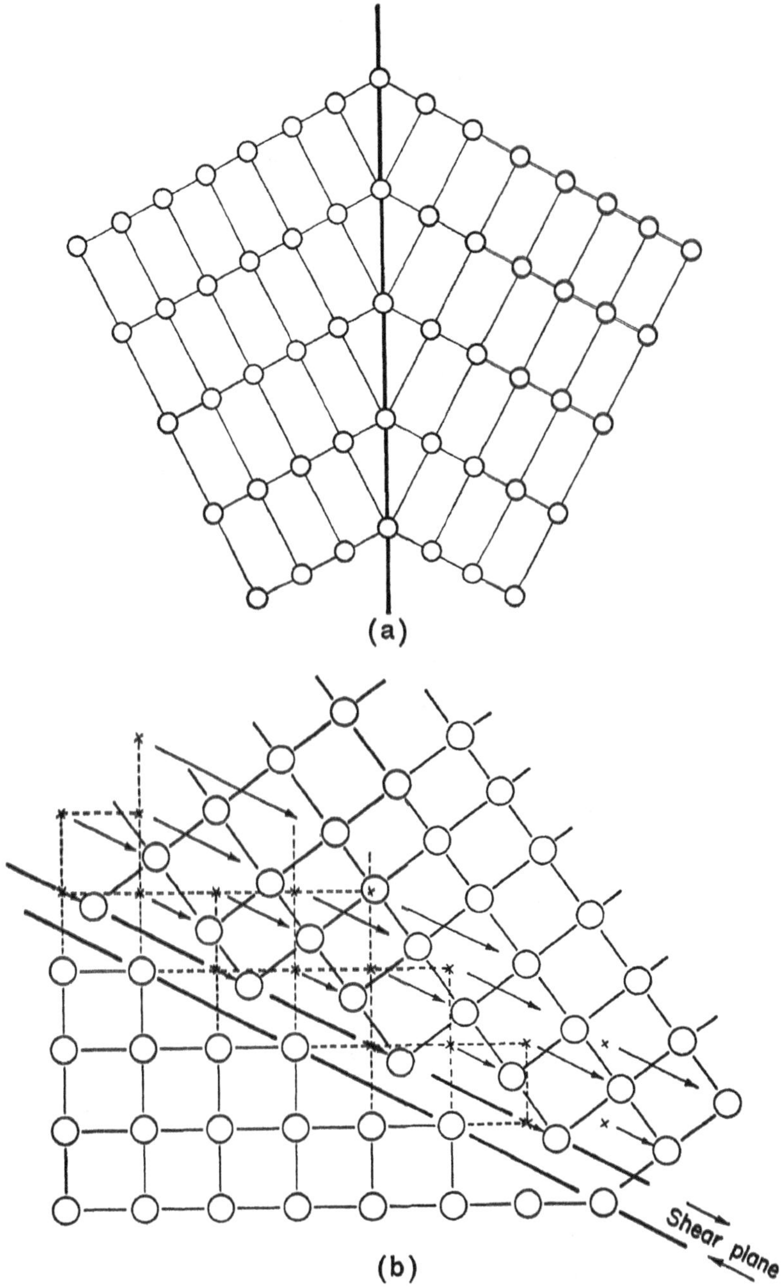

FIG. 3.15. (*a*) Twinning, showing the mirror-image positions. (*b*) Twinning by shear. The movements of the first few atoms above the shear plane from their original lattice positions of the twin are shown.

3.7. *The rhombohedral lattice*

In dealing with the close-packed structure, the difference between f.c.c. and c.p.h. lattices was seen to be the placing of layers C in the spaces between atoms in the B layer rather than directly over the A layer atoms (Figure 3.5). These close-packed layers form the octahedral planes (111) of the f.c.c. system, and can be illustrated in the f.c.c. unit cell by joining a corner atom to the nearest two face-centred atoms and the cube-face diagonal beyond these (Figure 3.16). The layer immediately below this layer, i.e. that obtained by joining the same atoms as before but starting at the diametrically opposed corner, will have its atoms displaced by 60° to those of the first

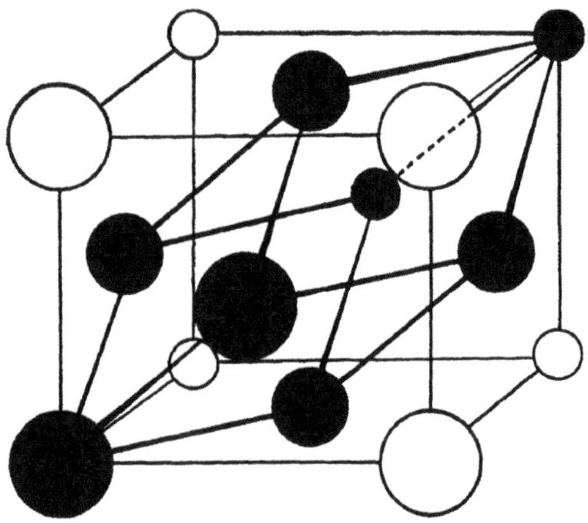

FIG. 3.16. The rhombohedral cell derived from an f.c.c. lattice.

layer—owing to the hexagonal symmetry. A unit cell can be constructed from these two planes as shown in Figure 3.16, in which the interplanar angles are 60°; each axis is equal in length. This is the *rhombohedral* cell of the close-packed system. Just as the hexagonal system need not be close-packed, so the rhombohedral cell need not have an interplanar angle of 60°. A number of materials crystallize in this way, including some metals. Mercury, for example, has a lattice angle of 70°, polonium one of 98°. Arsenic, antimony, and bismuth crystallize in the rhombohedral form, with a double-layer structure in which every atom has three near neighbours and three at a greater distance in the next layer.

The relationship between the close-packed structures and the rhombohedral structure does not have any further significance in the metals.

3.8. *The tetragonal lattice*

The basic tetragonal system is defined by three axes at right-angles, two of equal length. It resembles, therefore, the simple cubic system, in which the *c* axis has been altered to a c/a ratio greater or less than 1 but never equal to 1. A body-centred form can exist, as with the square symmetry planes of the simple cubic, i.e. by interpenetration of two unit cells. A face-centred form can also exist, and occurs with the metal indium. The c/a ratio for indium is so near unity (1·08) that the lattice is more nearly a face-centred cubic with slight distortion. White tin, i.e. the normal form, is tetragonal in structure. Since tin and indium contain 5*p* electrons, they might be expected to crystallize in the diamond-type structure with covalent bonding. Tin has a low-temperature modification (grey tin) which does indeed do this, but its normal form of white tin is the tetragonal one. It is believed that hybridization of the 5*s* and 5*p* electrons, necessary for covalent bonding (see page 35), does not occur here. Certainly it does not occur with thallium and lead, which follow indium and tin in the next period.

3.9. *The structure of compounds*

The crystal lattice of an ionic compound depends on the size of the ions and on their valency. It has already been shown that Coulomb forces are responsible for the binding, and exchange forces for the repulsion. The example of sodium chloride, quoted in Chapter 2, is a simple compound of composition AB, but the lattice size is affected by the large chlorine atoms. Since they are so large, only six can surround a sodium ion, hence the adoption of the simple cube structure. Compounds of AB type are not confined to this structure, however, and, when the ions are of similar size a higher coordination number is possible, as exists in caesium chloride CsCl (Figure 3.17), in which the number is 8. When the discrepancy between ionic sizes is even wider than in NaCl, the coordination number may drop to 4, as in zinc blende (zinc sulphide) (Figure 3.18). This structure is not unlike that of the diamond. Figure 3.19 shows the change-over from one type of structure to another at certain ratios of ionic radii, and the closeness of the packing that results.

Certain compounds exist that obey the valency rule applied to ionic compounds, but do not possess all the properties of ionic compounds. From the nature of the ionic bond, it is to be expected that a solvent such as water will not only dissolve them but also cause

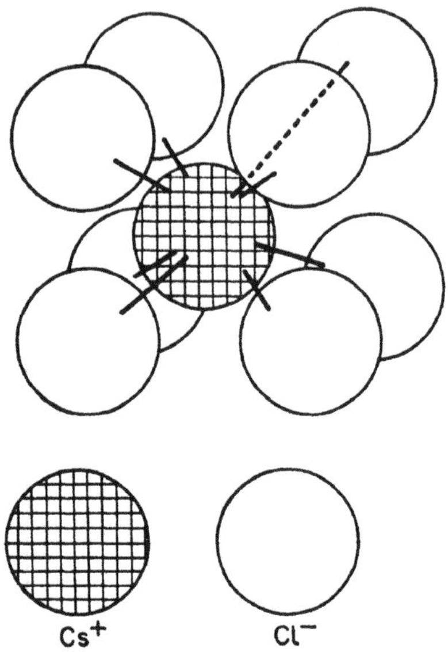

FIG. 3.17. The CsCl structure.

dissociation of the ions, since water itself is polarized (into H^+ and OH^- ions). When compounds are insoluble in water, and particularly when they show high hardness and melting point, it is suspected that the bond may be more covalent than ionic although there are excep-

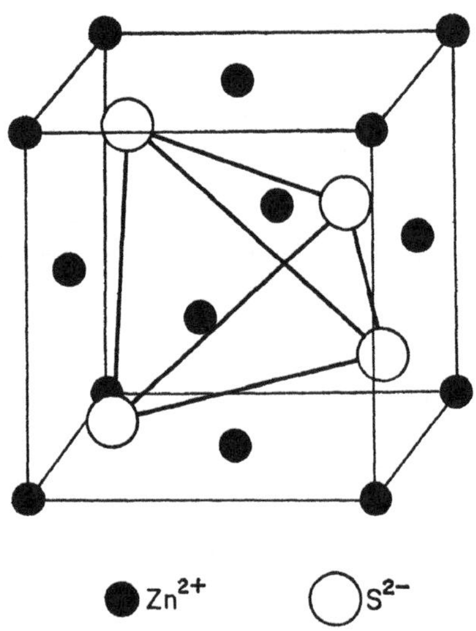

FIG. 3.18. The ZnS structure.

tions. The crystal lattice does not then obey the changes illustrated in Figure 3.19.

Examples of other ionic bond compounds of AB type (but mostly insoluble in water) are magnesium oxide MgO and many of the sulphides of divalent metals. The metallic compounds with selenium and tellurium (which behave like sulphur) are of AB type, e.g. MgSe, SnTe, and PbTe. Many of these are of interest because they are semiconducting.

Other examples of AB-type compounds that are insoluble in water, but of covalent bonding, are the silver halides, of great importance in photography (AgCl, AgBr, AgI). Silver iodide crystallizes in a

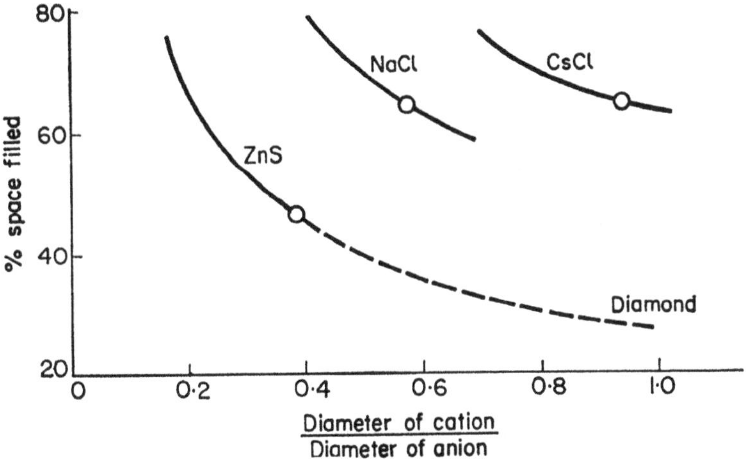

FIG. 3.19. Space-filling factor for compounds of AB type (after Zwikker).

diamond-type lattice, but the chloride and bromide have a higher coordination number than 4, and all contain largely covalent bonds. Compounds in the semiconductor class, with primarily covalent bonding, are InSb and GaAs. These contain four shared electrons per atom, made up of three from the group III member (In, Ga) and five from the group V member (Sb, As).

The next most common type of compound is that of composition AB_2. Halides of the alkaline earth metals tend to adopt the lattice of calcium fluoride CaF_2, commonly known as fluorite (or fluorspar). This is a face-centred cubic lattice of calcium ions, interpenetrated by a simple cubic lattice of fluorine ions (see Figure 3.20). The fluorine ions occupy the tetrahedral holes in the f.c.c. lattice, and so have a coordination number of 4. The coordination number of calcium ions is less than that in the normal f.c.c. lattice, by this same

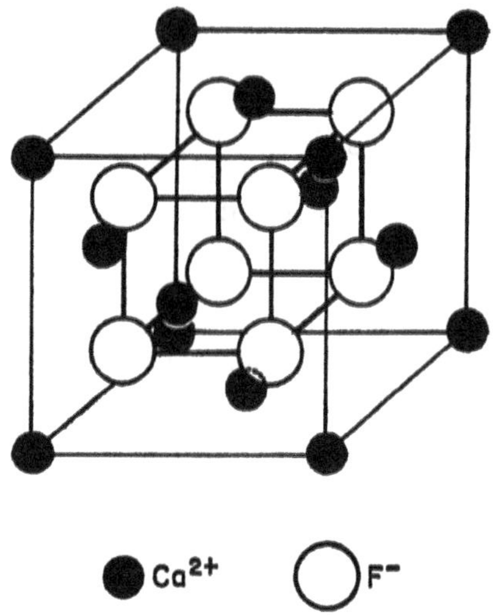

FIG. 3.20. The fluorite lattice CaF_2.

number, and so becomes 8 (8 + 4 = 12). The other types of AB_2 lattice are (i) that adopted by one form of titanium dioxide TiO_2, rutile (Figure 3.21) and (ii) that of cristobalite, an allotropic form of

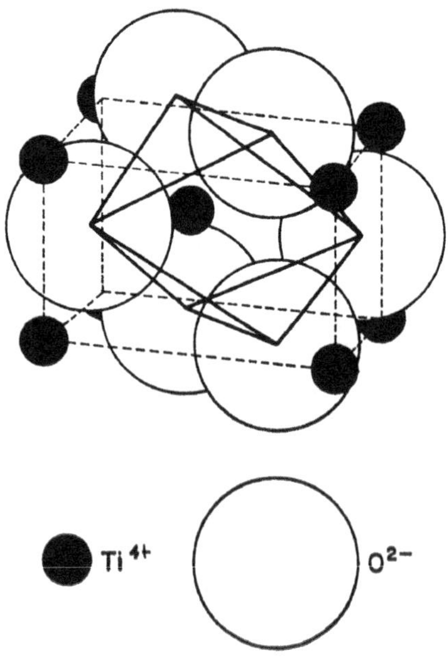

FIG. 3.21. The rutile lattice TiO_2.

silica SiO_2 (Figure 3.22). The coordination numbers in the rutile lattice are 3 and 6, but in the cristobalite type they are reduced to 2 and 4. Once again, the coordination numbers depend on the ratio of ionic sizes, and the space-filling factor is lowest for cristobalite. Many metallic oxides are of the AB_2 type, the majority being of the rutile structure. These lattices are all fairly 'open' in nature, and can depart from *stoichiometry*, i.e. strict composition ratios dependent on valency rules. The extra atoms may be metal or oxygen, provided, of course, that they can easily fit into the structure. Oxygen is the most common addition in the metallic oxide structures, and can dissolve in some lattices until saturation is reached at the next true

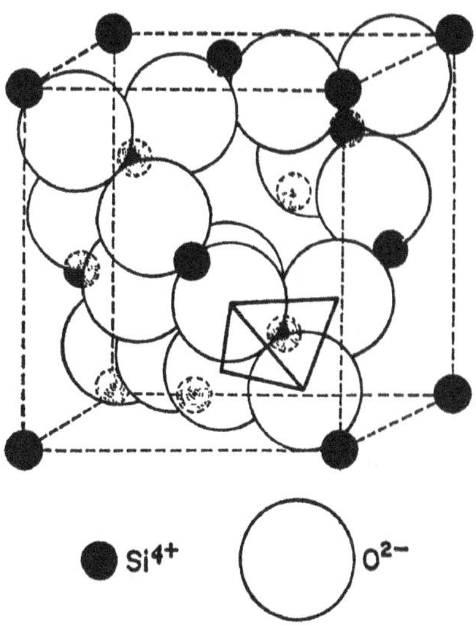

FIG. 3.22. The cristobalite lattice SiO_2.

valency composition and the lattice-type changes. Such compounds vary in composition, therefore, according to the oxygen pressure of their surroundings. In air, many of them slowly take up oxygen, whereas in a vacuum or an inert atmosphere they may remain stable. Heated in a stream of hydrogen, the excess oxygen may often be removed. These factors are important in the growing use of metallic oxides for high-temperature applications, since the properties commonly vary with composition, and the method of manufacture may be influenced by any tendency to depart from stoichiometry.

Figure 3.23 shows the effect of the ratio of ionic radii on structure and on the closeness of packing for AB_2 compounds. Cristobalite has entirely covalent bonds, and hence high hardness and high melting

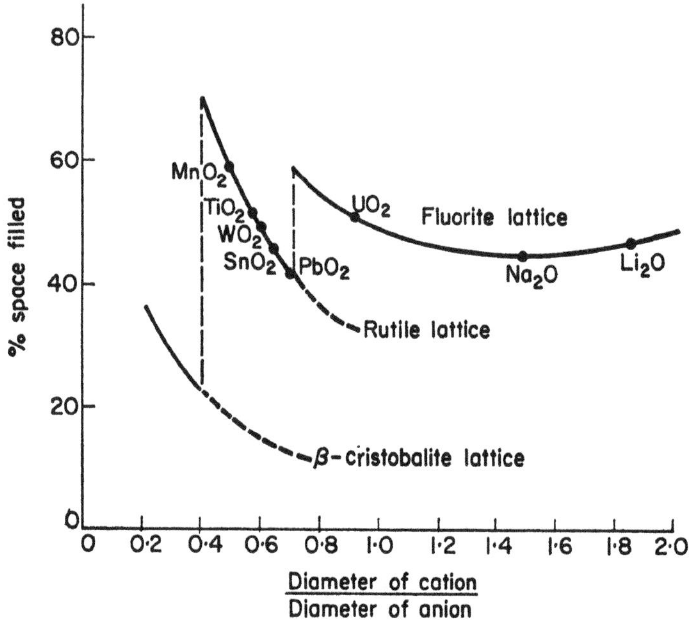

FIG. 3.23. Space-filling factor for compounds of AB_2 type (after Zwikker).

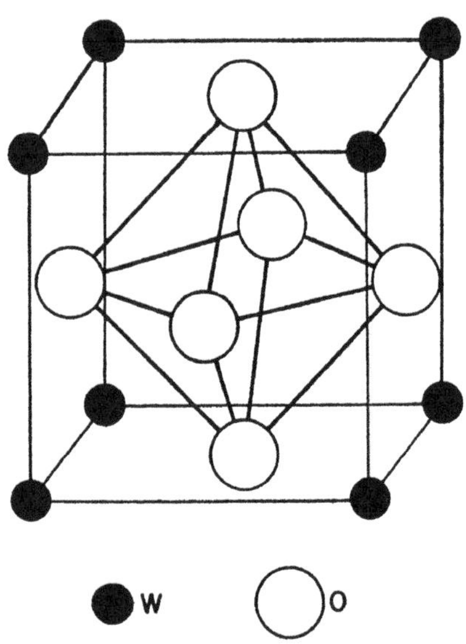

FIG. 3.24. The WO_3 lattice.

point. It is a constituent of refractory bricks for furnaces, being formed by high-temperature transformation from quartz (the low-temperature form of silica).

Many compounds of composition AB_3 have the lattice of tungsten trioxide WO_3 in which the tungsten atoms are at the corners of a simple cube and the oxygen atoms at the centre of each face (Figure 3.24). A similar type of lattice can exist in which a third small atom can occupy the body-centred position, i.e. the central octahedral void. An example of this is perovskite, $CaTiO_3$ (Figure 3.25). This

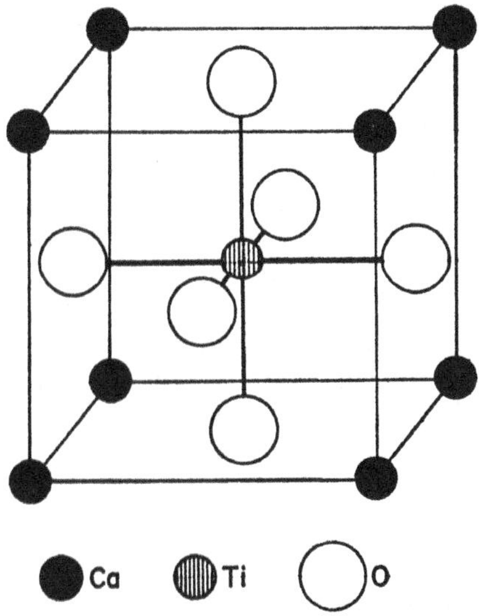

FIG. 3.25. The $CaTiO_3$ lattice (perovskite).

structure is significant, since the similar salt barium titanate $BaTiO_3$ is a *ferroelectric* crystal, i.e. one in which the permanent polarization of the crystal can be influenced by the application of an external electric field. Some crystals with the perovskite-type structure are also ferromagnetic. The compound WO_3 is ferroelectric.

More complex oxides exist, the most important being the *spinels*. They have the general formula AB_2O_4. The oxygen atoms are arranged in a cubic close-packing of 32 oxygen atoms to the unit cell. The metal atoms A and B are distributed in the tetrahedral voids (A) and octahedral voids (B) within this structure. When the A atoms are divalent metals (e.g. Mg) and the B atoms are trivalent (e.g. Al), then the structure is known as the *normal* spinel, after $MgAl_2O_4$, the first structure to be determined. A combination of metal atoms of valencies 2 and 4 is also possible, but the total charge on the metal ions must equal 8 to qualify for the spinel classification. The combination of trivalent atoms (Fe) in tetrahedral voids and divalent and trivalent atoms mixed (Mg and Fe) in the octahedral voids is often called an *inverse* spinel. Its formula is then best described in general terms as $B(AB)O_4$. Many compounds of this type exist, and their electrical and magnetic properties are of great importance. Magnetite, the magnetic oxide of iron, Fe_3O_4, is of this type, the iron having divalent and trivalent ions arranged in the inverse spinel structure, as $Fe^{3+}.(Fe^{+2}.Fe^{3+}).O_4$. Magnetite has high electrical conductivity, because of the exchange of electrons between Fe^{2+} and Fe^{3+}. Above 120°K an ordering of the ions in the octahedral positions causes an interruption in this electron exchange, and the conductivity decreases with temperature rise.

The spinels with fixed positions in the lattice are not conducting, and this fact is utilized in the so-called *ferrites*, spinel oxides containing trivalent iron, $Fe^{3+}(M^{2+}Fe^{+3})O_4$, M being a metal. These compounds are ferromagnetic, like magnetite, but the ferrites have a high resistivity, and eddy currents cannot be set up in them. They are used at frequencies in the microwave range, and also in computer devices for rapid switching because they are almost instantaneously saturated when magnetized. Only the inverse spinels are ferromagnetic.

The normal spinels are very stable oxides, and have quite close-packed structures. As a result, the diffusion of oxygen or metal ions through the lattice is very slow, even at high temperatures. This is a great advantage in heat-resisting applications, and the success of the alloys of nickel and chromium in this field is partly due to the formation of a pseudo-spinel oxide $NiCr_2O_4$ on the surface when exposed to oxygen. Such alloys allow some chromium atoms to be replaced by aluminium without modifying the spinel structure. The metal atoms in these pseudo-spinels are so disposed that they retain some coherency with the metallic lattice of the alloy. As a result, the physical properties and mechanical strength of the metal and the oxide are reasonably compatible, and the oxide does not break away from the underlying alloy when subjected to thermal stress. The protective film is therefore able to maintain its role for very long periods of time and over a wide range of high temperatures.

3.10. *The silicate structure*

Mention has already been made of the cristobalite structure as an example of an AB_2-type compound with a very 'open' lattice, owing to the small radius of the silicon ion compared with that of oxygen.

The natural form of silica, quartz, has the same type of structure, as does the other allotropic form, tridymite. The only variation in these different forms of silica is the arrangement of the tetrahedra comprised of $(SiO_4)^{4-}$ ions, i.e. one silica atom at the centre and four oxygen atoms at the corners of the tetrahedron. The silicon atoms are enclosed by the four oxygen atoms exactly as in the tetrahedral holes of a close-packed lattice. The spacing of the oxygen atoms is

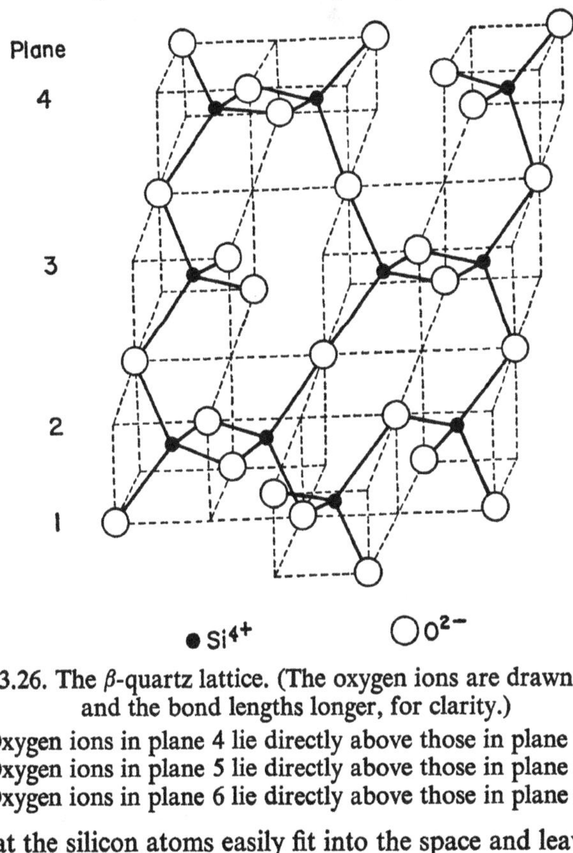

FIG. 3.26. The β-quartz lattice. (The oxygen ions are drawn smaller and the bond lengths longer, for clarity.)
Oxygen ions in plane 4 lie directly above those in plane 1.
Oxygen ions in plane 5 lie directly above those in plane 2.
Oxygen ions in plane 6 lie directly above those in plane 3, etc.

such that the silicon atoms easily fit into the space and leave a comparatively open structure. Each oxygen atom is shared between two tetrahedra; hence the composition of the ion (SiO_4) leads to a true chemical composition of one Si to $\frac{1}{2}$ O_4, i.e. SiO_2. Figure 3.26 illustrates the helical arrangement of the tetrahedra in β-quartz; Figure 3.27 shows that for tridymite.

The structure of silica, in its various allotropic forms, can thus be regarded as a network of oxygen ions, following the cubic (cristobalite) or hexagonal (tridymite) type of lattice, with silicon ions

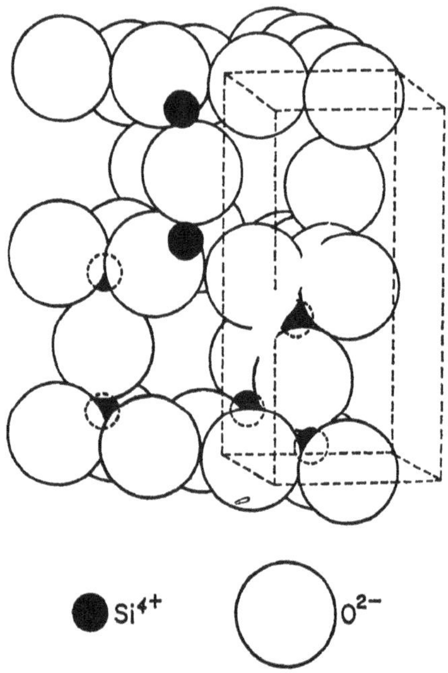

FIG. 3.27. The tridymite lattice SiO_2.

in the tetrahedral spaces. Quartz is a rather distorted form of the same structure, and at high temperatures quartz changes into one of the other two forms. There is still space in the crystal structure to accommodate small ions, especially in the octahedral spaces of the cubic or hexagonal types of lattice. There is also the possibility of the substitution of other ions for silicon, provided that the ion is small enough and that electrical neutrality is preserved by the balance between oxygen and metal ions. As a result of these possibilities, a whole range of substances, the silicates, can be built up. Many of these are natural substances. Silicon and oxygen are the most prevalent elements in the earth's crust, and the silicate group forms the basis of most rocks.

The size of the metal ions determines the composition of a silicate and affects the properties of the lattice. Beryllium ions are small enough to replace silicon easily; aluminium ions can do so with slight lattice distortion. Aluminium ions fit more readily into octahedral spaces in the lattice of close-packed oxygen ions, as do iron, magnesium, and titanium. These form silicate lattices with very little distortion, and hence fill up the spaces to make a more closely packed lattice. The larger metal ions, such as sodium, potassium, and calcium, cannot fit into the lattice without some distortion, and can only occupy octahedral spaces. These silicates are therefore less dense in the packing of their individual atoms than either pure silica or the group containing aluminium, iron, titanium, or magnesium. Figure 3.28 shows the approximate sizes of the relevant ions in the silicate structures, and Figure 3.29 gives an example of a silicate lattice.

The silica lattice is primarily formed by homopolar bonds, although the variation in size of the silicon and oxygen atoms must involve some ionic forces. The substitution of more strongly ionized

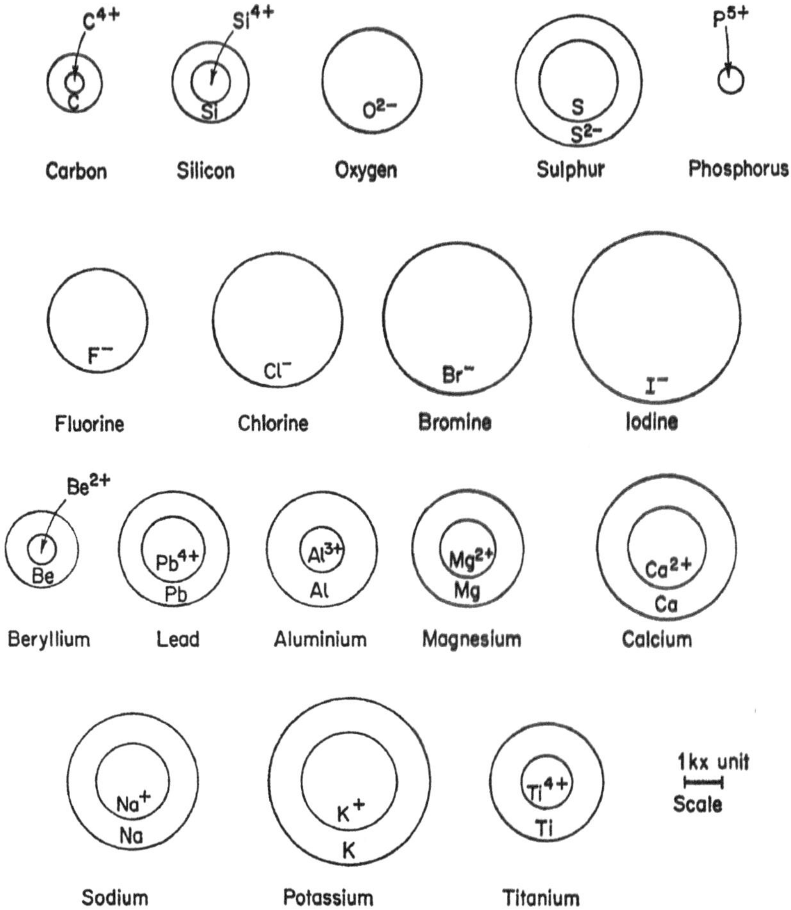

FIG. 3.28. The sizes of some representative atoms and their ions.
Ions of positive charge are smaller than their atoms.
Ions of negative charge are larger than their atoms.

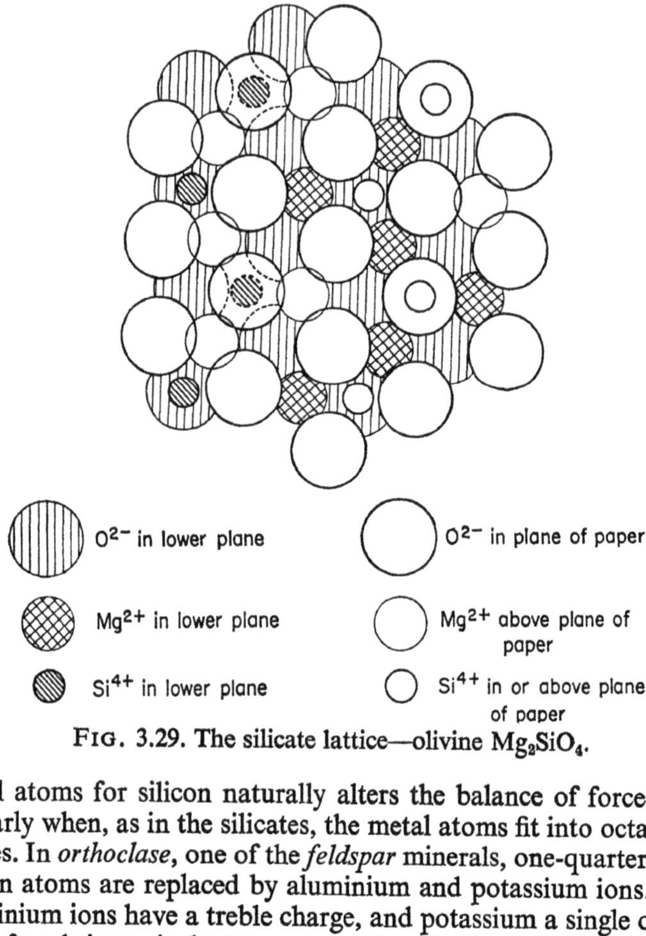

FIG. 3.29. The silicate lattice—olivine Mg_2SiO_4.

metal atoms for silicon naturally alters the balance of forces, particularly when, as in the silicates, the metal atoms fit into octahedral spaces. In *orthoclase*, one of the *feldspar* minerals, one-quarter of the silicon atoms are replaced by aluminium and potassium ions. Since aluminium ions have a treble charge, and potassium a single charge, one of each is equivalent to one quadruple-charged silicon ion. In the *mica* group, aluminium replaces silicon along with magnesium or iron; in garnet a trivalent metal (often Al) is combined with a divalent one (often Ca, Mg, or Fe^{2+}) to replace silicon. As electrical neutrality must be preserved, the proportion of silicon atoms replaced varies with the type of metal added. The density and close-packing, and therefore many properties, are accordingly altered.

Weathering of rocks produces some decomposition of the basic silicate minerals of the igneous rocks, and a new set of silicates results. In many cases the decomposition introduces water molecules into the structure, and a *layer* type of lattice, in which homopolar bonds are involved in the main structure, with some ionic forces, and Van der Waals' bonding between the layers. In other cases some ionic bonding between hydrogen and hydroxyl (OH^-) ions can occur. Thus, an entirely new set of properties follows. Layer lattices are strong in the plane of the layers but weak between them, since Van der Waals' forces are always weakest. A shear force readily separates the layers by plastic deformation, but compression does not produce gross deformation. Talc (Figure 3.30), a hydrated magnesium silicate,

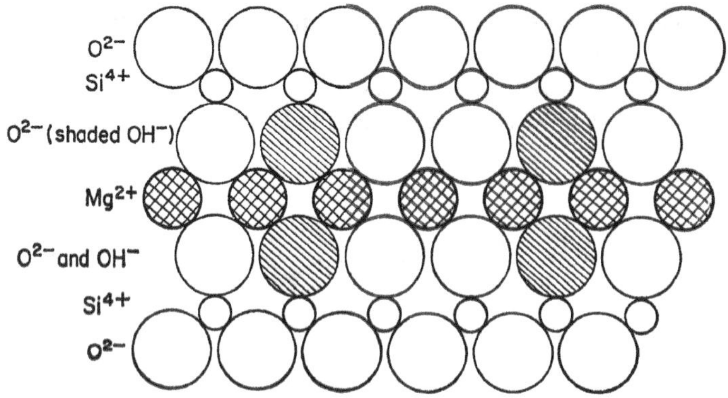

FIG. 3.30. Talc—a layer lattice.

possesses such a lattice, as does serpentine, in which iron is present in addition to magnesium. Weathering of feldspars produces the clay minerals, of which kaolin (china clay) is pure hydrated aluminium silicate (Figure 3.31). The clay minerals owe their plasticity to

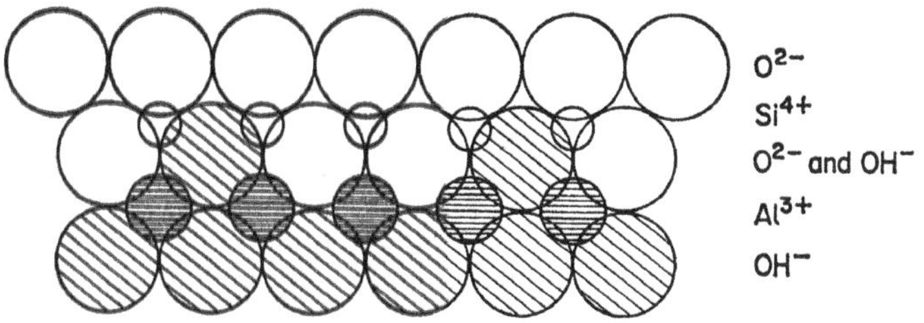

FIG. 3.31. The structure of kaolin.

the water of hydration, and the hydrogen–hydroxyl bonding that results. Once the clay mineral loses its water, as it does when fired at high temperature, the bonding is changed, and becomes amorphous (vitrified). The water cannot be replaced easily, and hence fired pottery remains reasonably strong but porous (owing to the holes left by the water).

When many lattices of the silica type are heated to a sufficiently high temperature and fuse together, they lose the crystalline state, and their long-range order changes to a very short range, i.e. the regularity of the lattice prevails over only a few molecules. This structure can also exist in the solid state, as when the fused mass is quickly cooled. This *amorphous* structure shows properties very similar to those of the crystalline one, since it is so similar. As it is metastable, it can be changed back into the crystalline state under certain conditions, such as the application of high temperatures for long periods of time. The application of stress may accelerate the process by assisting the movement of atoms into the positions of the regular lattice. Figure 3.32 shows the amorphous and crystalline forms of silica represented as a plane lattice. Vitreous silica is a valuable material, for it shows

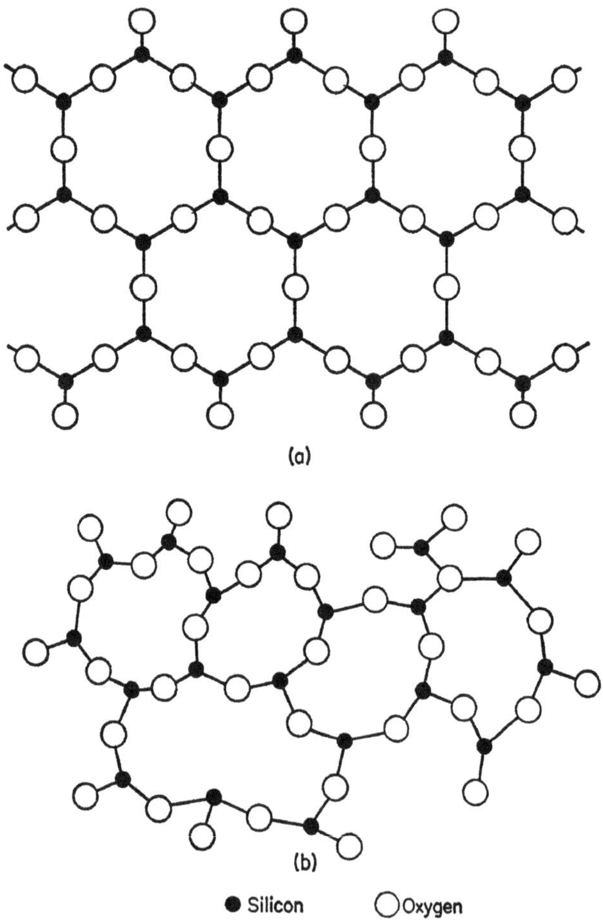

FIG. 3.32. (*a*) Crystalline and (*b*) vitreous (amorphous) silica.

a very low coefficient of thermal expansion. It also shows low thermal conductivity (as do most vitreous substances) but the thermal strains are low owing to the small expansion and so the material resists thermal shock. Fused silica is therefore extensively used in laboratories and industry for chemical plant, having great resistance to corrosion in addition to resistance to thermal shock. Fused silica is a *glass*, i.e. a material that has been so rapidly cooled from the liquid state that it has not had time to crystallize. It is therefore called a *supercooled liquid.* The liquid state of an apparent solid is best demonstrated by the fact that heating merely decreases the viscosity, or softens the material, which shows no well-defined melting point. On the other hand, crystalline quartz heated to the melting point melts at a well-defined temperature (which is dependent on its purity). *Slowly cooled,* quartz can again crystallize, as cristobalite, which is the high-temperature form.

Natural forms of amorphous silica occur in rocks that have been subjected to high temperatures and rapid cooling owing to igneous activity or earth movements. *Obsidian* is a natural silicate glass possessing the 'open' form of silicate network with short-range order. The cements and glasses are amorphous, as are many plastics.

3.11. *Glasses*

As mentioned above, any material possessing the properties of a supercooled liquid, having failed to crystallize and therefore being amorphous in structure, can be called a glass. The majority, but not all, are based on silicates; boron can also form borates which are glasses, phosphorus forms phosphate glasses, and aluminium the aluminates. The word *glass* relates simply to the amorphous structure of the supercooled liquid, and even some metallic oxides can be produced, at least partially, in this form by rapid cooling from the melt (see Figure 3.33).

The composition of glasses based on silicon, boron, and phosphorus depends on the same principles as those applied to crystalline compounds of these elements. The replacement of silicon ions in the silica tetrahedra by sodium, potassium, calcium, or lead can readily lead to the production of glasses. Figure 3.28 indicates the relative sizes of these metallic ions, all of which are larger than the silicon they replace. Although there is room to spare in the silica tetrahedral network, nevertheless the specific volume is increased, partly because of the larger metallic ions, but also due to the difference in binding energy between Si^{4+} and the electrically equivalent $4K^{+}$ or $2Ca^{2+}$.

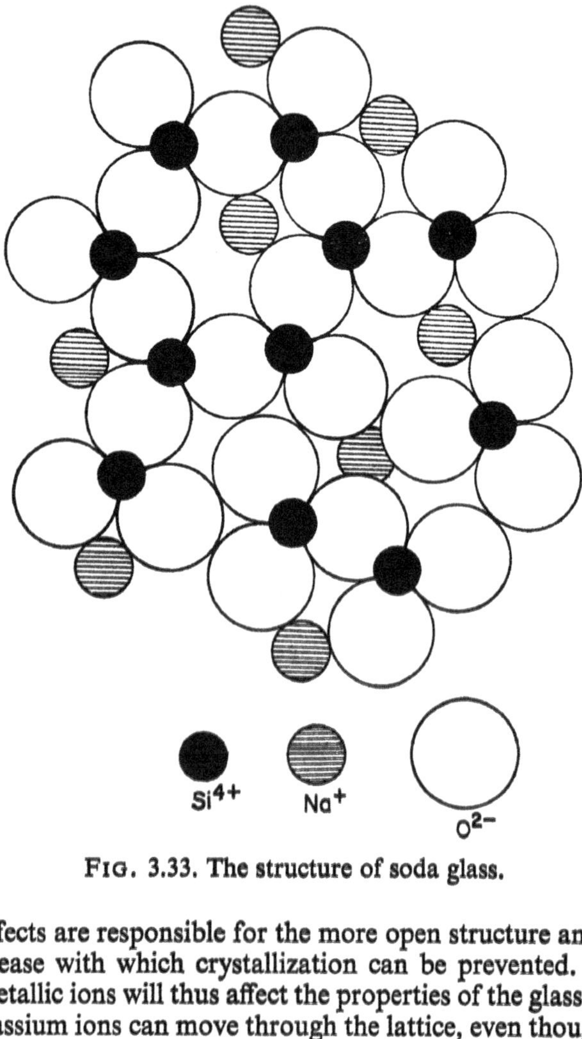

FIG. 3.33. The structure of soda glass.

These effects are responsible for the more open structure and for the relative ease with which crystallization can be prevented. The size of the metallic ions will thus affect the properties of the glass. Sodium and potassium ions can move through the lattice, even though amorphous, and a small degree of electrical conductivity results in glasses containing these. Sodium ions are smaller than potassium, and lead ions smaller than calcium, so that the softer glasses tend to be those incorporating sodium and lead, the intermediate ones contain some calcium, and the hardest potassium. Window glass, which is required to be rolled readily into sheets, contains sodium, together with lead or calcium. Glass for easy blowing into laboratory apparatus contains more lead, and the harder glasses for optical instruments more potassium.

The silicon ions may also be replaced, as in the crystalline silicates, by ions of smaller size, such as boron (B^{3+}), phosphorus (P^{5+}), or titanium (Ti^{4+}). Since boron and phosphorus differ in electrical charge (3 and 5 in place of 4) from silicon, the metallic ions must also vary in proportion when these additions are made, to preserve electrical neutrality. The boron and phosphorus thus form a new type of glass, usually referred to as *borate* or *phosphate glass*. The best properties usually result from the combination of silicon with boron *or* phosphorus, with the usual metallic ions. These elements are called *network-forming* ions.

In other cases, the metallic ions of intermediate size can replace silicon without the same strain on the lattice as is produced by Na^+, K^+, etc., and can either *modify* the network or assist in *forming* it. Typical ions in this category are Al^{3+}, Be^{2+}, Zn^{2+}, and Fe^{3+}. The first two are certainly small enough to replace silicon. Aluminium occurs in the natural glass obsidian, as a network former, in the proportion of one aluminium to five silicon ions, with sodium and potassium to modify the network.

When the ions added are of larger size than silicon, or of lower chemical valency (electric charge) the cohesion in the structure is reduced and so the softening temperature is lower. The resistance to chemical attack may be lowered; the mechanical strength most certainly is. A similar effect can be produced by replacing oxygen ions by halogen atoms, as in the fluor-silicate glasses. Again, the effect is due to the lower valency.

The colour of glass is affected by the type of ions present. The addition of cobalt can make the glass blue or pink, according to whether cobalt acts as a network former or modifier, respectively. Heating glass containing metallic ions that are network-modifying tends to move them into positions that are network-forming and so varies the colour. Rapid strides are continually being made in the development of glasses with 'tailor-made' properties.

3.12. *Carbon compounds*

In Chapter 2 the covalent type of bond was discussed with reference to the diamond lattice of carbon. The structures of silicates, depending as shown above on an ionic lattice of tetrahedra, resemble the diamond lattice in structural configuration, but in little else. The structures of carbon compounds stem directly from the covalent or homopolar linkage, and the numerous permutations and combinations that this makes possible.

As explained in the previous chapter (page 42) carbon atoms in the form of diamond achieve a stable octet of electrons by four bonds, each comprised of a pair of shared electrons. Each of these pairs shows two electrons of opposite spin, achieving all the necessary requirements to satisfy the Pauli principle and quantum and wave mechanics. The hybridized sp^3 state of these electrons can be regarded as an overlapping band between $2s$ and $2p$ states, holding four electrons per atom. No more electrons may enter this band, and a considerable energy gap exists before the next permitted band; therefore diamond is an insulator.

Another form of carbon, graphite, can exist, in which the bonding

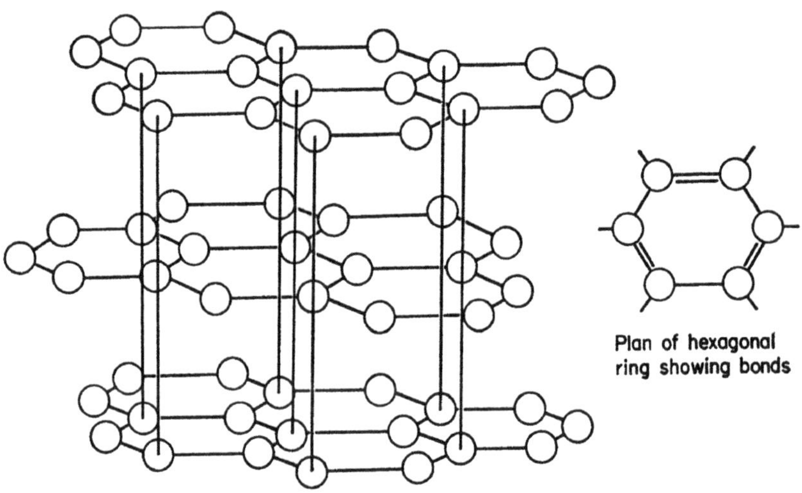

FIG. 3.34. The graphite lattice and double bonds.

between atoms is in the form of a 'ring' structure. If six carbon atoms form a hexagonal 'ring', then each atom has three nearest neighbours, ond cannot satisfy the stable octet except by creating a fourth bond. This it achieves by making one bond double, i.e. sharing four electrons with one other atom. The same structure applies in certain carbon compounds, e.g. benzene. The hexagonal rings can extend in two dimensions over a very large area, forming a 'layer' lattice. Between the layers, Van der Waals' attraction results from the polarized nature of the carbon rings. The double bonds, of which there must be three per hexagon (see Figure 3.34), are not fixed, but 'resonate' around the ring, so that all the four bonds of any one atom will be double bonds for one-quarter of their existence. It is not possible to say exactly where the bonds are at any one time, any more than it is possible to say where conduction electrons are in the metallic lattice, but obviously the Pauli principles and the wave-mechanical requirements can be met by such an arrangement. The Van der Waals' forces are of course much weaker than the homopolar bonds; graphite is therefore readily cleaved into layers, making it a good lubricant, but it has considerable strength in the plane of the layers of hexagonal grouped atoms. Owing to the resonance of the double bonds, electrons can be affected by an electric field, and graphite is a conductor, but only in the direction of the hexagonal rings. In a solid lump of graphite a fairly random distribution of the layer lattice can result in reasonable conductivity, but when it is extruded into a shape such that the layers tend to lie in one particular direction there will be marked anisotropy, as discussed with reference to the hexagonal metal structure. Graphite thus combines the homopolar bond (in its layer structure) with Van der Waals' attraction between the layers, producing unusual properties.

A third, amorphous, form of carbon also exists. This can be produced only under specific conditions, e.g. as soot from the decomposition of certain carbon compounds such as benzene. Here there is incomplete combustion, and groups of carbon atoms rather like graphite hexagons can be sublimed on to a cold surface, i.e. prevented from crystallizing. There is a very short-range order in the carbon structure, but it behaves as an amorphous material. There is less anisotropy than in graphite and great porosity, and thus less reflectivity at the surface. Lamp-black, produced by this decomposition and sublimation, is used for paint, since the pigment is a poor reflector and produces a dense black.

The compounds of carbon arise from the formation of covalent bonds with other elements. The introduction of hydrogen leads to a build-up of several types of hydrocarbon, either in the form of chains or as rings; in both cases there is opportunity for branching or cross-linking. The chain compounds are mostly the *aliphatic* compounds, i.e. the main constituents of many 'fatty' or oily substances. The ring compounds are called 'aromatics' because they include the substances having well-defined aromas—many being used for flavouring in foods.

The aliphatic compounds are built on a structure of four bonds of shared electron pairs. Some of these may be double or even triple bonds, with resultant lower bond strength owing to the distortion of the directional effect. Acetylenes, with triple bonds, readily ignite and may be explosive. Not only hydrogen may substitute for carbon, but any grouping with electrons to share in appropriate proportions. The combination of oxygen with hydrogen and carbon leads to a wide variety of substances: the alcohols, the aldehydes and ketones, the fatty acids, etc. The branching of the carbon chains gives rise to 'isomers', substances with the same atomic proportions but different structures and with some difference in properties. The polarization of the molecule is changed by this branching, and consequently properties affected by Van der Waals' attraction, e.g. melting and boiling points,

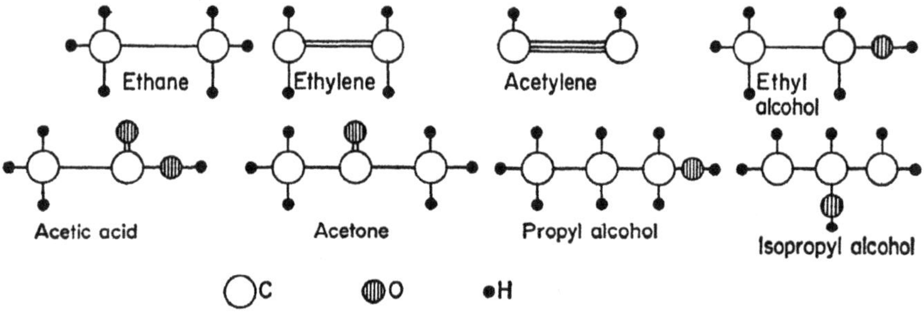

FIG. 3.35. The structures of some carbon chain compounds (not to scale).

are immediately affected. Figure 3.35 illustrates the structures of some chain compounds.

The aromatic compounds are derived from the 'benzene' ring, which can build up to form complex compounds that form the basis of many industrial and biological processes. Some of these are illustrated in Figure 3.36. The 'resonance' of double and single bonds

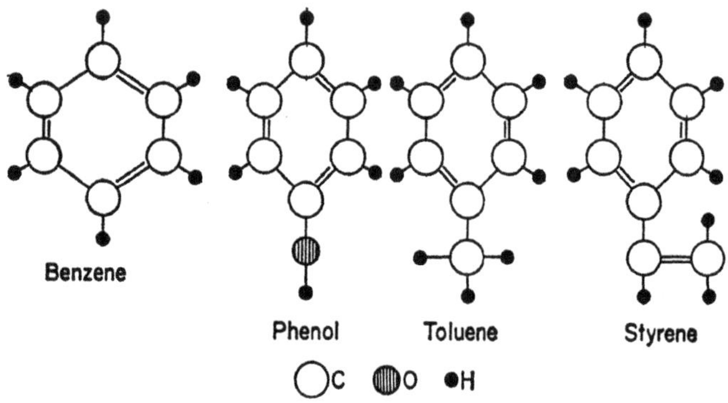

FIG. 3.36. The structures of some carbon ring compounds (not to scale).

referred to in connection with graphite again applies, unless all are saturated by the substitutional atoms.

The basic unit from which a compound of the chain type can be built up is a $—CH_2—$ group, i.e. one carbon atom with two hydrogen atoms and two as yet unfilled bonds. These combine with similar units in the *'paraffins'*, of general formula C_nH_{2n+2}, but have double bonds in *olefines* (C_nH_{2n}) and triple bonds in *acetylenes* (C_nH_{2n-2}). The last two groups of compounds are said to be 'unsaturated' because they contain multiple bonds of shared electrons. Many aromatic compounds are unsaturated for the same reason, as with benzene (C_6H_6). It is possible to build up these basic units into very large molecules indeed, and many of these occur naturally, as in petroleum. By forcing the basic units together, e.g. under high pressure, some of the unsaturated bonds may be broken and transformed into two single bonds; union of several such units may cause a very large molecule to build up again. The basic unit is called a *mer*, the normal small molecule of a similar unit which can combine to form a big one a *monomer*, and the built-up large molecule a *polymer*.

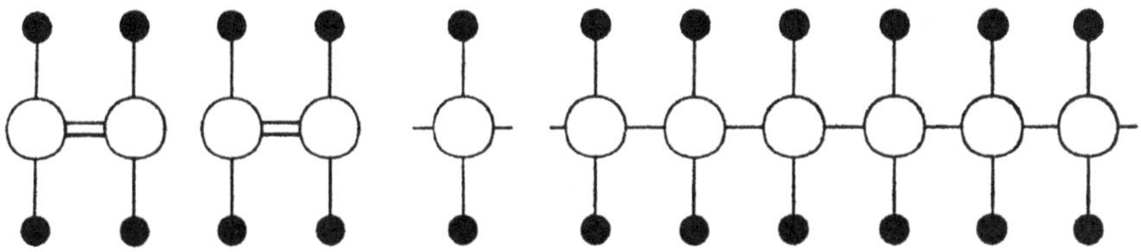

FIG. 3.37. Structure of polyethylene.

Figure 3.37 indicates the meaning of these terms with respect to polyethylene. The final structure is obviously a paraffin hydrocarbon with a very long chain.

If the saturated hydrocarbons are heated to a high temperature, they break up into smaller unsaturated molecules, the reverse of polymerization, known as 'cracking'. This is done in the petroleum industry, to utilize what would otherwise be wasted residues of very viscous high boiling point hydrocarbons. After cracking, the unsaturated hydrocarbons are treated with hydrogen in the presence of a catalyst, to substitute hydrogen at the double or triple bonds and convert the unsaturated molecule to a saturated one. The structures of these compounds are identical whether natural or synthetic.

Polymerization is now a well-established technique for building up molecules that may be difficult to separate from natural products, or are not normally found in nature. The properties can be largely tailor-made by appropriate treatments, and will be discussed later. From the structural aspect, there is one more important point. The polarization that results from these large molecules will depend on the atoms present and their position. Electric dipoles may occur along the length of a chain molecule or at particular positions in a ring compound. These dipoles tend to align themselves together between individual molecules, and hence produce Van der Waals' attraction in appropriate cases. Thus, not only are we concerned with the bond strength in large molecules but also with forces between these. The more entangled the individual molecules become, the stronger may be the material. The method of production of these polymers has a profound influence on the orientation of the molecules, quite apart from the efficiency of the polymerization process itself. In the *elastomers*, a rotation of single carbon atom bonds occurs so that the long chains become spirals or even approach spheres. When subjected to tension, the molecule unwinds, but returns to its former shape if the stress is not too high.

Cross-linking, i.e. the joining together of two molecules at specific points, may also be caused by the presence of other chemical substances. The vulcanizing of rubber is nothing more than cross-linking to produce a larger, more complex, molecule with obviously changed properties. The properties that result from these structural changes are discussed in more detail in Chapter 9.

4

Imperfections in Crystals

The ideal crystal structures referred to in Chapter 3 are very rarely achieved. Although the vast majority of atoms are situated as described by a given form of crystal structure, there are imperfections of various kinds present to greater or lesser degree in all substances. It is only comparatively recently that these defects have been recognized and their role in the behaviour of materials appreciated.

4.1. *The surface*

The surfaces of solid substances are never perfectly plane or smooth. The closest packing of atoms possible would automatically prevent this, in view of the space remaining between atoms as described in Chapter 3. In addition, the methods adopted for producing materials, by solidification from a melt, pressing, or other methods of shaping

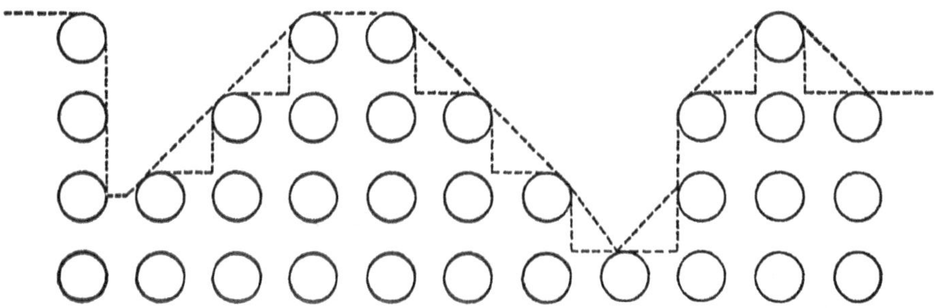

FIG. 4.1. Irregularities of structure in the surface (dotted lines represent feasible contours).

and machining, cannot give a perfect surface for they involve an already imperfect surface of mould or tool. Atoms will therefore be present in surface irregularities of large or small clusters, following the crystal structure, but separated along the surface by gaps in the structure. Figure 4.1 illustrates this point.

4.2. *Vacant lattice sites*

On a given surface of normal crystal structure a further type of defect may arise by the migration of an atom from below the surface layer of atoms. This stems from the continual thermal vibration of atoms at temperatures above absolute zero, which increases in magnitude with rise in temperature. The migrating atom leaves behind a hole in the crystal lattice, known therefore as a *vacant lattice site* or *vacancy*. This type of defect is known as a *Schottky defect*, after the inventor of the term. Owing to the continual vibration, the migrating atom may return to the lattice and move through it to annihilate the original vacancy, or some other vacancy. This can happen only if a drop in energy occurs as a result. Such surface imperfections may therefore be continually changing, but they are, of course, on an atomic scale, and become most important at high temperatures, e.g. in thermionic emission and oxidation processes.

The Schottky defect (see Figure 4.2) is not limited to the migration of an atom to the surface, but refers to the production of a vacancy whenever a migrating atom moves to some position that does not

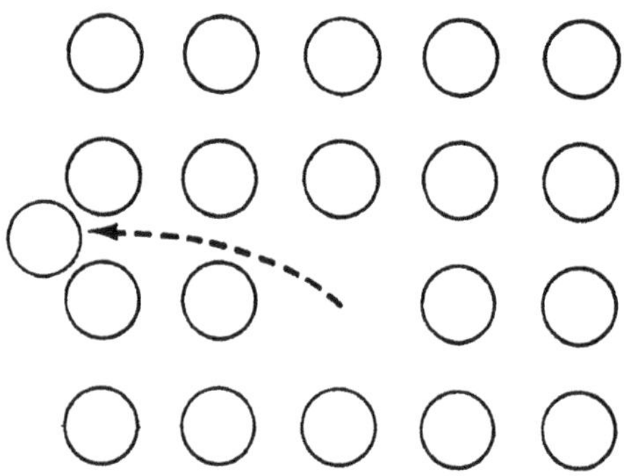

FIG. 4.2. Schottky defects.

create disturbance in the remaining lattice, i.e. the surface of a void or other sink of disordered atoms.

Vacant lattice sites are also formed by the migrating atom occupying an *interstitial* position in the lattice (an octahedral or tetrahedral hole in the case of cubic lattices) and are then known as *Frenkel defects*, after the man who first described them (Figure 4.3). Vacancies and interstitials are often associated together, and both can

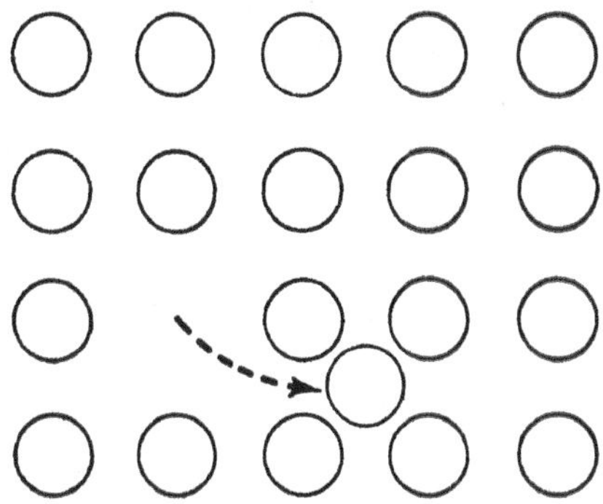

FIG. 4.3. Frenkel defects.

associate with like atoms or holes in a cluster. A large association of vacancies is obviously a void, and a large cluster of interstitials must severely strain the lattice. Thermal activation can cause movement of the vacancy or interstitial whether associated together as a pair, or with other atoms or holes in a cluster. These vacancies or interstitials thus diffuse readily through a lattice. In ionic crystals, either Schottky or Frenkel defects would create an imbalance of electric charge if only one ion were involved, and there is therefore a strong tendency for pairs of ions to move. In sodium chloride and similar AB-type structures, an ion pair can readily diffuse and so preserve electrical neutrality. The vacant lattice sites are then positive and negative ion sites, present in equal number. The electrolytic conductivity of such salts is obviously affected by this diffusion of ion pairs.

The effect of temperature on thermal vibration, and hence on the production and diffusion of these defects, can be illustrated by examples. The ratio of vacancies n to the number of atoms N is given approximately by

$$\frac{n}{N} \simeq e^{-W/kT}$$

where W is the work done to remove an atom from the lattice to a sink, and T the absolute temperature. The value of W for copper is of the order of 10^{-19} joules, or 1 electron volt (eV). This gives a vacancy distribution of about 0·001% at 1,000°K. For ion pairs forming positive and negative vacancies, as in sodium chloride,

$$\frac{n}{N} \simeq e^{-W_p/2kT}$$

where W_p is the energy of formation of the pair. The value of W_p for sodium chloride is approximately 2 eV, so that at room temperature there is one defect ion pair for every 10^{16} ions. At the melting point, some of the silver salts, such as silver bromide, show defect concentrations of the order of 0·01%. The heat capacity of silver bromide is accordingly affected by this large number of defects, and increases by a factor of 3 between 500°C and the melting point at 700°C.

Owing to the temperature effect, any material very rapidly cooled from a high temperature (e.g. metals quenched as in heat treatment) contains more than the equilibrium number of vacancies and interstitials at room temperature. Diffusion processes are thus affected on reheating by the number of holes and interstitial atoms that are not in equilibrium and therefore have excess free energy.

4.3. *Interstitial atoms*

Quite apart from the interstitial atoms resulting from the creation of a Frenkel defect, and similar processes to be described later, there may be interstitial atoms present in a lattice as impurities. The addition of a second element to a substance can result in *substitution* of the solute for the solvent atom, or the packing of the solute into an interstitial position. There are appropriate conditions for substitutional atoms, which will be referred to later. With interstitials, the size of the available holes in the lattice is important, and was referred to in Chapter 3. An impurity may be trapped in such a position even when of unfavourable size, but cannot be retained in any quantity, as it would be when in *solid solution* with the appropriate size factor. Nevertheless, the presence of an interstitial impurity causes an interruption of the periodicity of the lattice, and can thus affect many properties. It may diffuse through the lattice if small enough, and may interact with vacancies during thermal vibration or diffusion. The strain on the lattice and the effect on properties can justify the inclusion of such atoms as imperfections.

Vacancies, interstitial atoms and impurities are referred to as *point defects* for obvious reasons.

4.4. *Line and plane defects*

The regularity of the bulk crystal is not necessarily perfect, for reasons similar to those given for the defects in the surface. The crystallization of solids from a melt is governed by many factors, notably the rate and direction of cooling, the formation or presence of nuclei from which crystals can grow, and the rate of growth. More than one nucleus inevitably gives rise to a *polycrystal*, where several points of growth lead to crystallites of different crystallographic orientation. These subsequently interfere with each other to form *grain boundaries*. The boundaries give rise to lines of mismatched lattice, where adjacent atoms may sometimes be closer to each other, but are generally farther away from each other, than in the ideal lattice

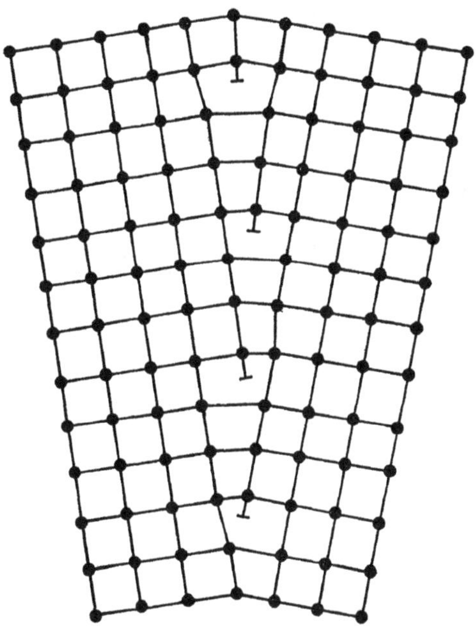

FIG. 4.4. A simple grain boundary.

(Figure 4.4). Similarly, growth within the crystallites may be affected by vacant lattice sites, interstitial atoms, or impurities. When atoms are inserted between normal planes over parts of the crystal only, a host of imperfections may result, according to the subsequent thermal and mechanical history of the crystallite. Strain is produced in the lattice as a result of these line or plane imperfections, considerably modifying the physical and mechanical behaviour of the material. These imperfections are called *dislocations* because of their influence on the lattice.

4.5. *Dislocations*

Most substances deform by the gliding of atom planes; the plane and direction depending on the type of crystal structure. Calculations on the force required to cause such movements give results far above the experimentally determined values. In the early 1930s, several workers sought to explain these discrepancies, notably Orowan, Polanyi, and Taylor. TAYLOR advanced the hypothesis of an extra half-plane of atoms, inserted between two parallel planes of a normal crystal, which would easily distort the atoms in the immediate vicinity. Slip would take place by movement of this extra half-plane to left or right, with much less force than would normally be necessary. He called this imperfection an *edge dislocation*. Figure 4.5 shows

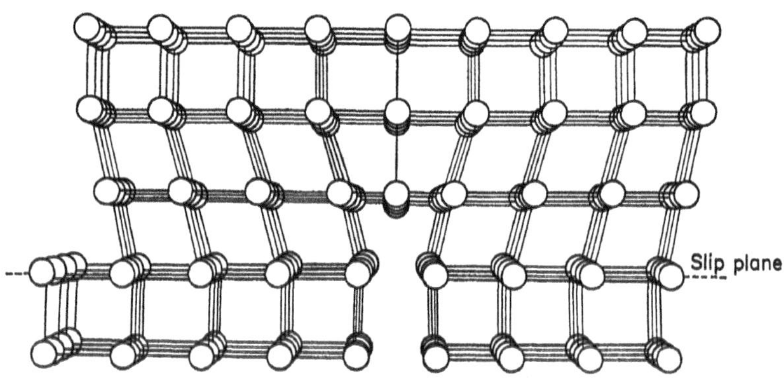

FIG. 4.5. An edge dislocation (after Goldman).

the extra half-plane of atoms, and the result of slip in causing dislocation movement. MOTT has used the analogy of a carpet being dragged across a roughened floor to illustrate the process of slip. A heavy carpet and rough floor would require a considerable force to move the whole area of carpet simultaneously—analogous to slip over a whole plane of atoms in a crystal. If a fold is produced in the carpet first (see Figure 4.6), then movement of the fold in either direction will be much easier, but eventually the result of this slip will effectively cover the whole area of carpet. The fold is analagous to an edge dislocation. Deformation in crystals is therefore dependent on the number of dislocations present and their distribution. An edge dislocation itself moves, although the atoms of both regular and dislocated lattice only move a distance equal to one-half the normal distance apart, as can be seen from Figure 4.5. In the edge dislocation, slip takes place along a plane perpendicular to the dis-

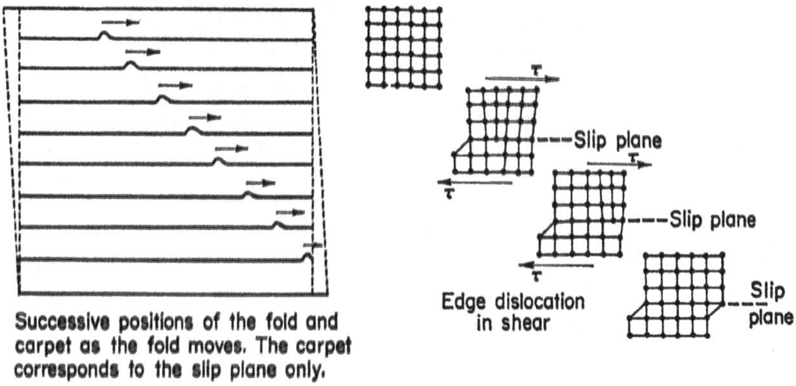

FIG. 4.6. Movement of an edge dislocation in shear, with the analogy of a carpet.

location itself, and the slip plane and slip direction remain those of the particular lattice type involved. Such deformation requires a considerable area of the slip plane to move, and this is not necessarily true in practice. There are localized areas of deformation even in one crystal grain. BURGERS proposed a second type of dislocation, in which the displacement of atoms is not confined to a half-plane parallel to the normal planes of the lattice, but is arranged on two separate planes perpendicular to each other, in a helical fashion. This is known as a *screw dislocation* for this reason, and is a little

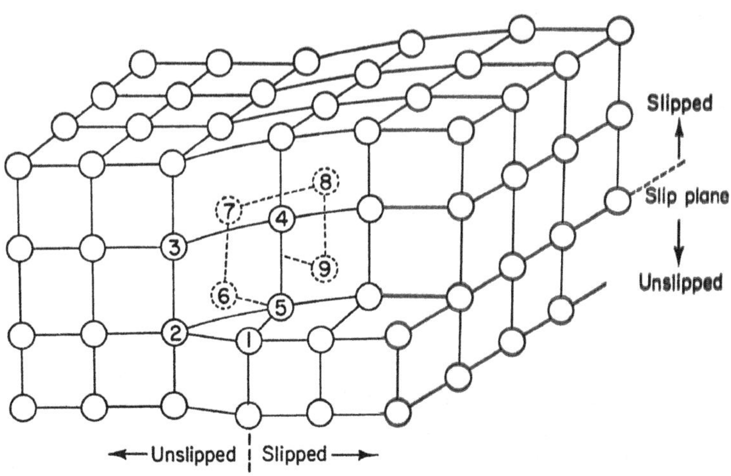

FIG. 4.7. A screw dislocation in a cubic lattice. (The numbered atoms indicate the helical part of the screw.)

more difficult to envisage. Figure 4.7 illustrates this type of imperfection as a three-dimensional model. In the screw dislocation, the glide direction is parallel to the dislocation, in contrast to the perpendicular direction for the edge type. When a crystal grain contains elements of both edge and screw dislocations, a well-defined area of slip results, often in the form of a regular figure called a *dislocation loop*. The edge components are perpendicular to the slip direction and the screw ones parallel to the slip. The motion of a screw dislocation is not confined to a single slip plane: when it moves into another slip plane having a slip direction common to the original one this is called *cross-slip*.

The displacement of the half-crystals meeting on the slip plane when a dislocation moves is given the symbol b and called the *Burgers vector*, a characteristic of the dislocation. If a shear stress τ acts on a slip plane of area A, an element of dislocation $\mathrm{d}l$ will move through a distance $\mathrm{d}\varepsilon$, and the area swept by the movement is $\mathrm{d}l\mathrm{d}\varepsilon$. The mean displacement of the half-crystals is then $(\mathrm{d}l\mathrm{d}\varepsilon/A)b$.

The applied force is $A\tau$, so that the work done in slip is given by

$$A\tau(\mathrm{d}l\mathrm{d}\varepsilon/\mathrm{A})b = \tau b\mathrm{d}\varepsilon\mathrm{d}l$$

The force acting on the dislocation, F_d, causes the movement $\mathrm{d}\varepsilon$,

$$\text{Work done during slip} = F_\mathrm{d}\mathrm{d}\varepsilon$$

and

$$F_\mathrm{d} = \tau b\mathrm{d}l$$

acting on length $\mathrm{d}l$.

The force per unit length of dislocation must therefore be τb, directed towards the unslipped portion of the crystal.

In well-annealed polycrystalline metals, the dislocation density is of the order of 10^5 to 10^8 cm^{-2}. Heavily deformed metals show an increase in dislocation density to 10^{11} or 10^{12} cm^{-2}. Calculation of the strength of metals, taking into account this dislocation density and the effect on strength, reduces the theoretical values by factors of 10^2 to 10^4, which is of the order of the experimental values.

4.6. *Dislocation climb*

Although the normal movement of dislocations is along a slip plane, other movements are possible. Screw dislocations can only move by slip because they are oriented parallel to the slip plane. Edge dislocations, however, having the extra atoms perpendicular to the slip plane, may move in the direction of the extra half-plane itself. If a vacant lattice site migrates to the end of the dislocation, then the dislocation moves up or down by one atom spacing at that point. If an interstitial atom migrates to the dislocation end, the same thing occurs, but the extra half-plane is of course now longer than it was. This phenomenon, movement in the direction of the extra half-plane of atoms of an edge dislocation, is called *climb* (Figure 4.8). Dislocations can surmount obstacles in their path, by climbing into another slip plane and proceeding thence by normal slip. The process

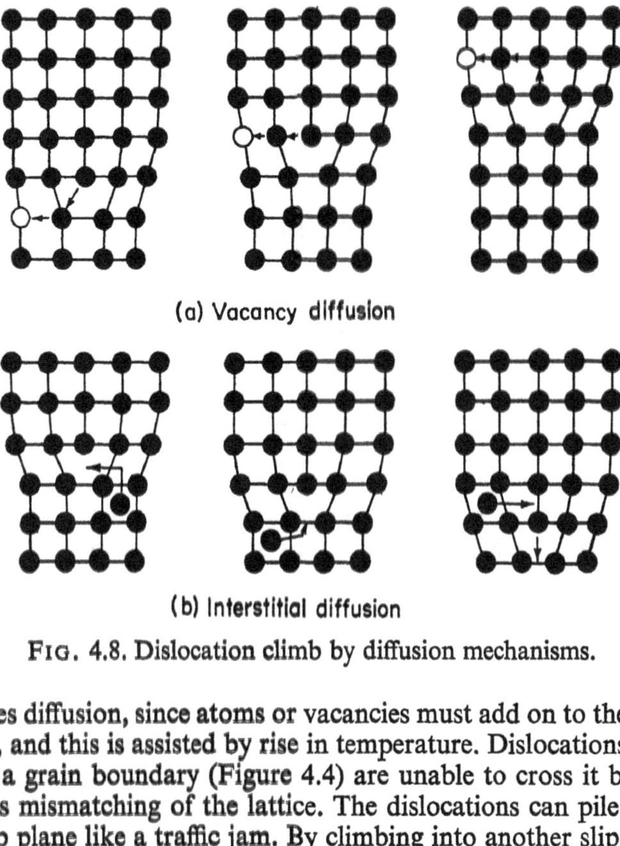

FIG. 4.8. Dislocation climb by diffusion mechanisms.

involves diffusion, since atoms or vacancies must add on to the dislocation, and this is assisted by rise in temperature. Dislocations arriving at a grain boundary (Figure 4.4) are unable to cross it because there is mismatching of the lattice. The dislocations can pile up on the slip plane like a traffic jam. By climbing into another slip plane, or set of planes, the strain induced by the pile-up at a grain boundary is reduced. This process arises in the recovery of metals by annealing after deformation—often producing a network and known as *polygonization.* Figure 4.9 illustrates polygonization after a pile-up of dislocations.

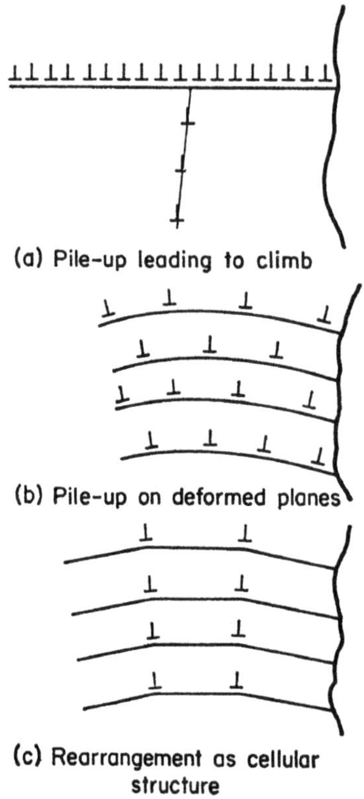

FIG. 4.9. Polygonization as a result of dislocation climb.

4.7. *Dislocation jogs*

As can be seen from a consideration of dislocation climb, the addition or subtraction of atoms from an edge dislocation need not be uniform across the half-plane. In this case, the dislocation may operate on more than one slip plane when viewed three-dimensionally, one or more steps occurring between the various operative slip planes. These steps are known as *jogs*. Figure 4.10 illustrates a jog in an edge dislocation. In these cases, the jog can move with the dislocation by the normal process of slip. Screw dislocations may also have jogs, but since the slip direction is that of the dislocation line, the jog resembles a small piece of edge dislocation, and cannot move so easily. If the jog climbs, by moving into another slip plane, point defects are created, but the screw dislocation containing the jog can continue to slip. Otherwise the jog must move sideways along the

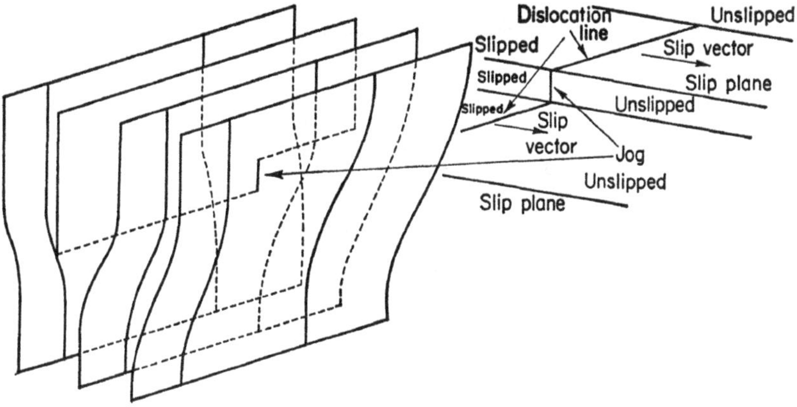

FIG. 4.10. A dislocation jog.

screw dislocation and attach itself to an edge component of the dislocation line. These movements are more difficult, and are thought to be involved in the work-hardening of metals.

4.8. *Imperfect dislocations*

The Burgers vector of a dislocation defines the displacement of atoms resulting from dislocation glide or slip. When this is equal to a whole-number multiple of the lattice spacing in the crystal, the slipped regions are restored to perfect matching of the crystal planes after the dislocation has passed; such dislocations are therefore called *perfect*. It is not essential, however, that a whole-number multiple of lattice spacing should be involved; indeed, in some crystal structures the movement of dislocations is more difficult by such means. Dissociation of a dislocation into two portions that move together as a unit, but in which each portion slips by only a fraction of lattice spacing, can occur; the dislocations are then termed *imperfect* or *partial* dislocations. Partial dislocations are also involved in the process of twinning, and reference back to Figure 3.15 will show atomic movements in twinning that are only a fraction of the lattice spacing. In such cases, the crystal lattice between partial dislocations is disordered, for movements involving a fraction of lattice spacing are implied. This disordered lattice is known as a *stacking fault*. In the case of two partial dislocations forming a unit, the stacking fault is removed after the passage of the second partial dislocation through the slip plane (Figure 4.11). Such partial dislocations are known as *Shockley* partials. The width of the stacking fault, i.e. the distance apart of the partials, depends on the stacking-fault energy of the

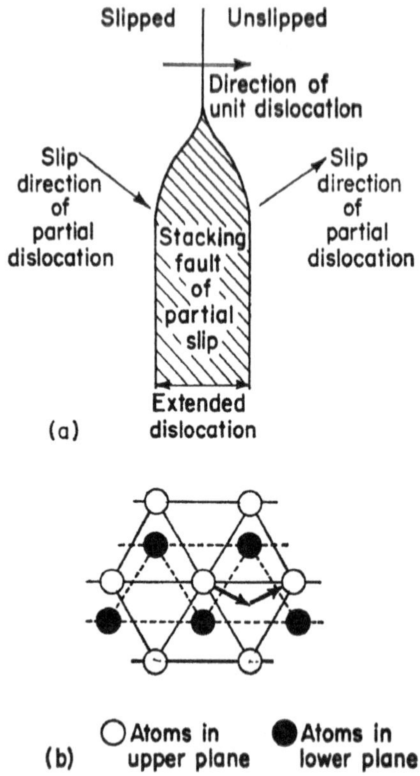

FIG. 4.11. (*a*) An extended dislocation. (*b*) One possible partial atom movement involved.

particular crystal. The higher this energy, the narrower the strip of fault. Some materials are therefore more prone to form stacking faults than others. Metals such as copper, silver, and gold have low values of this stacking-fault energy, and so readily form partial dislocations with strips of stacking fault. Aluminium and magnesium have high values and so areas of stacking fault are limited. SMALLMAN likens stacking-fault ribbons to a thin twin crystal, and twin crystals are rarely seen in aluminium. When zinc is alloyed with copper, the lattice remains face-centred cubic, the stacking-fault energy is lowered from the value applicable to pure copper. As a result, twins produced by annealing deformed alpha-brass are more common than in copper subjected to the same treatment.

In copper, having low-stacking fault energy, the partial dislocations are about eight atoms apart, whereas in aluminium they are only one or two atoms apart. These dislocations are often called *extended* dislocations in view of the separation of two partials, which, with the ribbon of stacking fault, constitute a unit moving together across the slip plane.

4.9. *Sessile dislocations*

Dislocations that can move are often called *glissile* to distinguish them from those that are pinned down, called *sessile*. FRANK described one form of sessile dislocation, in which an aggregate of vacant lattice sites collapses to form a loop of dislocation surrounding a disk of stacking fault, as shown by Figure 4.12. The stacking

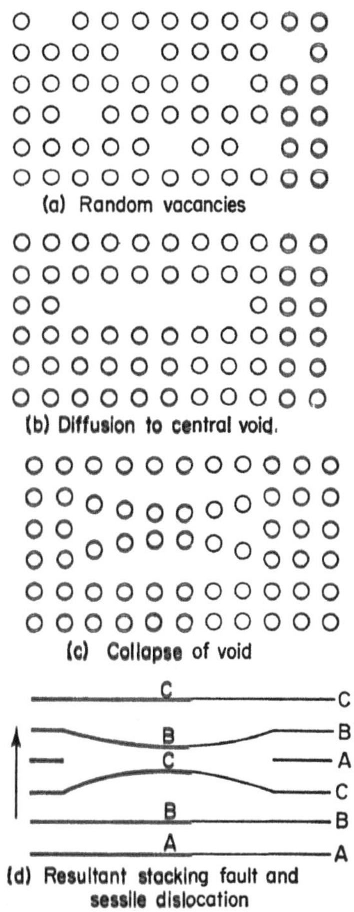

FIG. 4.12. Frank sessile dislocation.

fault is here readily seen by the change in stacking sequence of planes. Similar faults can be demonstrated for hexagonal and body-centred cubic lattices.

4.10. *Frank–Read sources*

The increase in dislocation density in crystals as a result of deformation was referred to in Section 4.5. This increase cannot be accounted for by thermal vibration of atoms producing misfits in the lattice at odd places, and since slip is localized on a limited number of planes the increase in dislocation density is even more puzzling. One explanation for the increase invokes the function of a continuously operating source of dislocations, first described by F. C. FRANK and W. T. READ. If a dislocation is fixed at one end by being situated on

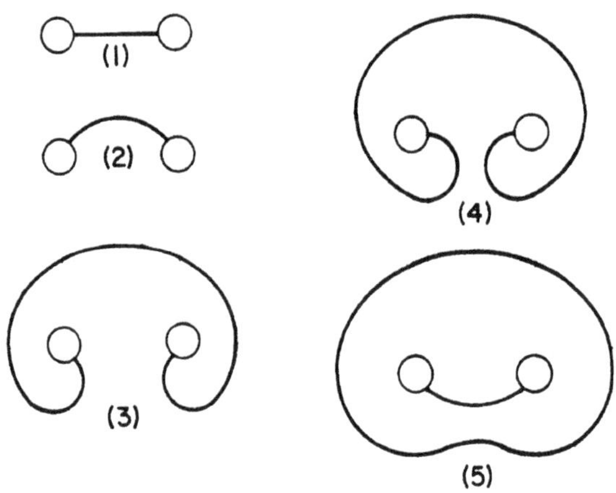

FIG. 4.13. The stages involved in operation of the Frank–Read source.

an inactive slip plane, the dislocation line can rotate about that point, and every rotation along the slip plane produces slip corresponding to the Burgers vector. Continual rotation under stress would therefore produce unlimited slip so long as the stress remained. This would not account in itself for the increase in dislocation density.

If both ends of the dislocation line are pinned, however, but the remainder of the line is in an active slip plane, the process described above can proceed with the formation of loops of dislocation. The bulging line loops round and leaves an expanding loop plus a restoration of the original line, and the process starts again (Figure 4.13). Continual slip on the active plane is therefore produced, with loops of dislocation spreading out from the source. This explains why the dislocation density increases, and why slip is limited to certain active planes.

Direct observation of such sources has been reported from electron micrographs, but as yet the evidence is scanty. Nevertheless, some mechanism of this kind is essential to explain the facts on plastic deformation.

5

The Vibration of Atoms and their Thermal Properties

5.1. *The lattice vibration*

Atoms in a crystal lattice are obviously not stationary. The equilibrium position taken up as a result of the opposing forces of attraction and repulsion is unlikely to remain undisturbed. Any input of thermal energy is immediately made manifest as an increase in the mobility of the atoms, and therefore an increase in vibration about the mean position. The capacity of the material to absorb this energy is called the *specific heat*, and the increased mobility is manifested as *thermal expansion*. The transmission of the energy through a mass of material is measured as *thermal conductivity*. All three parameters are closely linked, since they stem from the same atomic disturbance. There are at least two other important parameters linked to these atomic vibrations, which will be dealt with later, those referring to elasticity and electrical conductivity.

5.2. *Specific heat*

The equation of state of a perfect gas may be written

$$pV = RT$$

where RT represents the energy of one gramme-molecule at temperature T and R, the universal gas constant, is 1·988 cal degK^{-1}.

When referring to a single molecule, the equation is written

$$pV = kT$$

where k, Boltzmann's constant, is $1{\cdot}380 \times 10^{-16}$ erg degK^{-1}.

To measure the specific heat, or heat capacity, defined as the quantity of thermal energy absorbed by unit mass when temperature rises by one degree, we must keep either pressure or volume constant. The specific heat at constant volume is given by

$$C_v = dE/dT$$

where dE is the increment of internal energy absorbed when temperature rises by an increment dT. The atoms in a monatomic gas have three degrees of vibrational freedom, so that their average energy is $(3/2)kT$ per atom. Multiplying by N_0, Avogadro's number, the number of atoms in a gramme-atom, we have the energy for a mass of gas, $N_0(3/2)kT = (3/2)RT(N_0k = R)$.

Then
$$C_v = \frac{d(3RT/2)}{dT} = \frac{3}{2}R \tag{5.1}$$

In a crystal, however, the atoms are bound together rather differently, and can absorb thermal energy only by vibrating about their mean positions in the lattice. In order that the whole system may remain unchanged, the atoms must all vibrate with the same frequency. In a crystal of N atoms, there will be $3N$ degrees of freedom (three mutually perpendicular directions) and therefore $3N$ normal modes of vibration. The simplest approach to a physical picture of this vibration is that of three harmonic oscillators vibrating along three mutually perpendicular directions, not in one single mode, but a superposition of the $3N$ possible frequencies. The average kinetic energy of a classical harmonic oscillator is $\frac{1}{2}kT$. The potential energy is also $\frac{1}{2}kT$, and the total average energy is thus kT per degree of freedom. The average energy per gramme-atom is thus $3N_0kT$, and

$$C_v = \frac{dE}{dT} = \frac{d}{dT}(3N_0kT) = 3N_0k = 3R \tag{5.2}$$

This is known as the law of DULONG and PETIT, who measured the specific heat of many solid substances and found an approach to the constant value $3R$ at elevated temperatures. Unfortunately, the specific heats of crystals tend to zero as the temperature is lowered, whereas the Dulong and Petit law implies no temperature dependence. The approach of classical mechanics must therefore be replaced by a quantum-mechanical consideration. A quantum harmonic oscillator can possess energy states given by

$$E_n = (m + \tfrac{1}{2})hn \tag{5.3}$$

where $m = 0, 1, 2, 3 \ldots$ The lowest energy state, when $m = 0$, still gives $E = \frac{1}{2}hn$; this means that at absolute zero there is a small but finite energy in the atoms of a crystal. The energy can only be varied in steps, or quanta, exactly as discussed in Chapter 1 with regard to the emission of spectral lines. The term *phonon* is given to the quantum of thermal energy in a similar manner to the term *photon* for the quantum of electromagnetic radiation.

As the temperature is raised from absolute zero, the lowest frequencies of vibration come into play first. The wavelengths of these vibrations will be large compared with the distance apart of the atoms. An increase of temperature will lead to progressively higher frequencies of vibration, and therefore a greater contribution of these frequencies to the specific heat. It is difficult to imagine a true physical picture of this complex vibration, but a root mean square of amplitude can be calculated to show the variations with temperature. In sodium, for instance, the values are approximately 0·2 Å at absolute zero, and 0·5 Å at room temperature. The atomic bonding in sodium is comparatively weak, and these figures are higher than for elements of stronger bonding.

Einstein first applied the quantum-mechanical approach to the theory of specific heats. He retained the assumption that each atom could vibrate independently of the others, and would have the same frequency. In addition he used a value of m in equation (5.3) $1/[\exp(hn/kT) - 1]$. The average vibrational energy for 1 gramme of material becomes

$$E_{av} = \frac{3N_0hn}{\exp(hn/kT)-1} \tag{5.4}$$

Differentiating with respect to temperature to obtain specific heat at any temperature

$$C_v = 3R\left(\frac{hn}{kT}\right)^2 \frac{\exp(hn/kT)}{[\exp(hn/kT) - 1]^2} \tag{5.5}$$

At very high temperatures, kT is very much greater than hn, and we can expand equation (5·4) to show that $E_{av} = 3N_0kT = 3RT$, the law of Dulong and Petit.

At very low temperatures, kT is much less than hn, and $\exp(hn/kT)$ is much greater than 1, so that the average energy will be

$$E_{av} = \frac{3N_0hn}{\exp(hn/kT)}$$

and

$$C_v = \frac{3N_0k(hn/kT)^2}{\exp(hn/kT)} \tag{5.6}$$

According to equation (5.6), therefore, the specific heat should decrease exponentially with decreasing temperature. For a given material, a characteristic frequency n_E can be found to give fair agreement with experimental measurement, and this can be used to define the Einstein characteristic temperature, $T_E = hn_E/k$. The characteristic frequencies n_E are of the order of 10^{12} to 10^{13} cycles per second (c/s), but the decrease in specific heat with fall in temperature is less than predicted by Einstein, and closely approaches proportionality to T^3. The Einstein model is thus too simple. DEBYE performed more sophisticated calculations, assuming that vibrations were not independent, and therefore there is no limit to the frequencies. To simplify the calculations, he assumed a cut-off point N_D at which the number of possible frequencies equals $3N$, although the total number of frequencies is actually made up in a different form. He then defined a characteristic temperature in a similar way to that used by Einstein, in this case known as the *Debye characteristic temperature* and normally written Θ_D, where

$$\Theta_D = hn_D/k \tag{5.7}$$

The specific heats of solids are then roughly equal at temperatures that are the same fraction of Θ_D and a plot of C_v against the ratio T/Θ_D shows how the points for many solids lie on the same curve. At low temperatures, the value of C_v should vary as T^3 from the Debye theory, and this is true of many substances. At high temperatures, the value of C_v should approach $3R$, which it does. The Debye theory is only an approximation, and does not take into account any contribution from the electron cloud. The specific heat due to free electrons is proportional to the absolute temperature, and is very small at room temperature. At very low temperatures, however, when lattice vibrations are tending to zero, the electronic specific heat is much more significant. Measurement of specific heat at low temperatures has accordingly been used to explore the electronic configuration of some solids, since the proportionality is dependent on the distribution of electrons in the highest energy levels.

The Debye characteristic temperature is constant for a given material, and is high when the binding forces in the atom are strong, and when the atomic weight is low. Typical values are 1,840°K for diamond, 159°K for sodium, and 315°K for copper. Figure 5.1a shows the variation of specific heat with temperature for two metals and for diamond. Figure 5.1b shows the same results plotted against the ratio of temperature to the Debye characteristic temperature.

In determinations of specific heat, it is normally the value C_p for constant pressure that is obtained, for obvious reasons. The value of C_v can be obtained using a correction,

$$C_p - C_v = TVK(3\alpha)^2$$

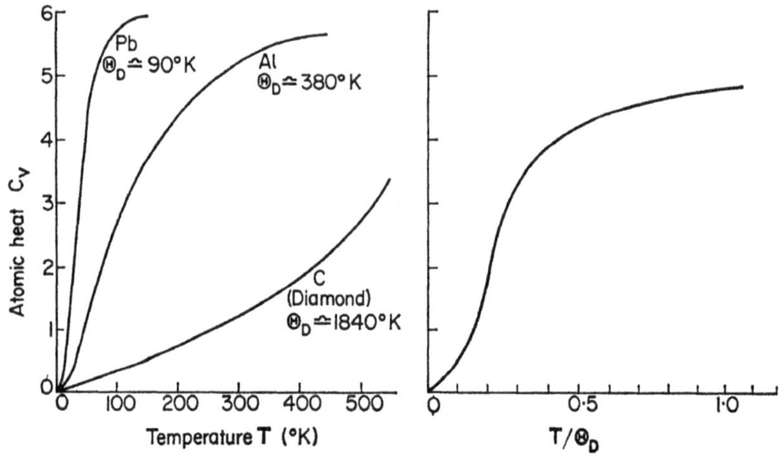

FIG. 5.1. (*a*) Variation of atomic heat with temperature. (*b*) Variation of atomic heat of a metal with ratio of absolute temperature to Debye characteristic temperature.

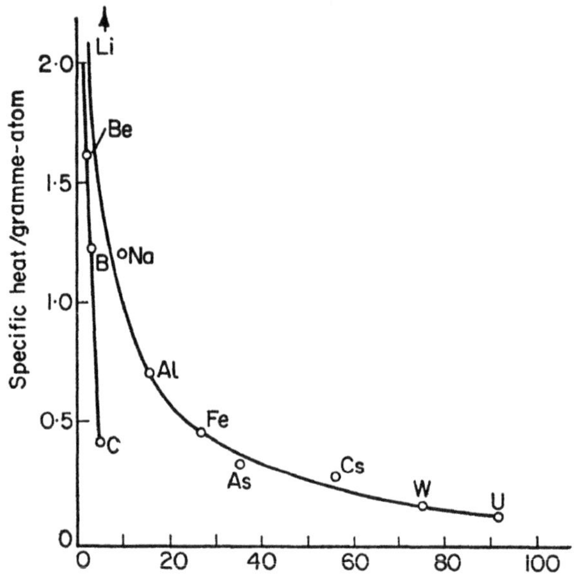

FIG. 5.2. Variation of specific heat per gramme-atom with atomic number.

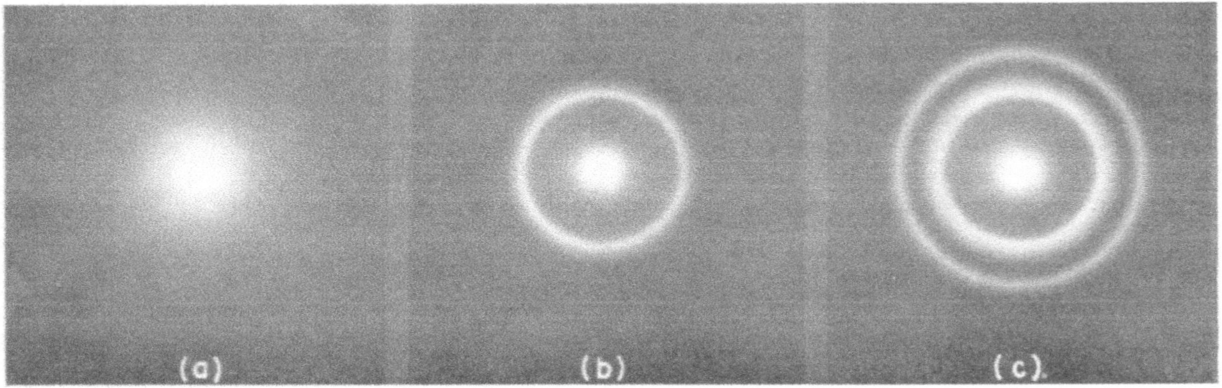

PLATE I. Simulation of electron density in hydrogen for the states (*a*) 1*s*, (*b*) 2*s*, (*c*) 3*s*.

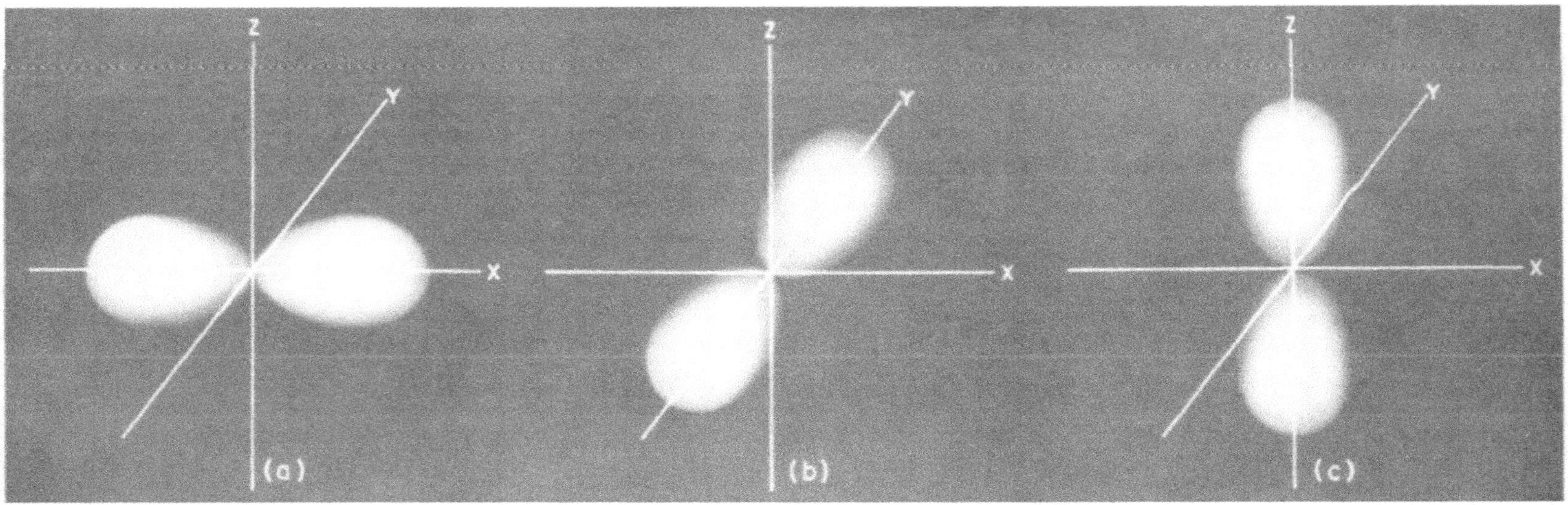

PLATE II. Simulation of the 2*p* electron state for a hydrogen-like atom. The three figures indicate the directional effect of an applied field. (*a*) $m_l = +1$; (*b*) $m_l = -1$; (*c*) $m_l = 0$.

PLATE III. Crystal lattice of NaCl.

PLATE IV. A close-packed plane of spheres.

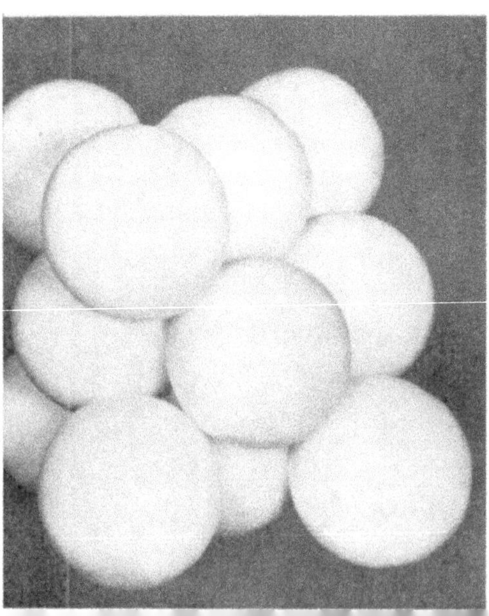

PLATE V. An octahedral plane in a face-centred cubic lattice.

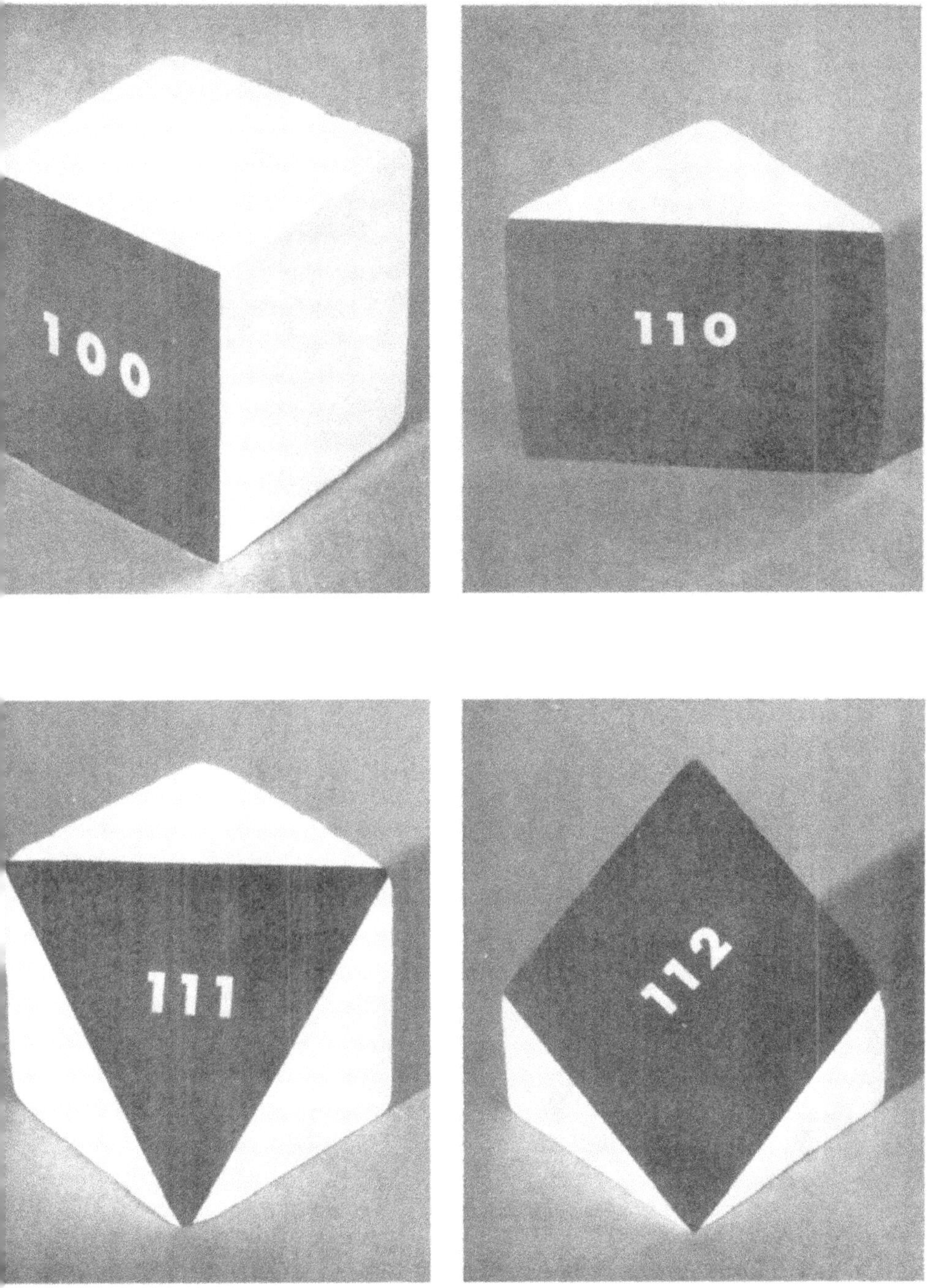

PLATE VI. Miller indices for some cubic planes.

PLATE VII. Production of a twin crystal by rotation.

where T is the temperature, V the volume per gramme-atom, K the bulk modulus of elasticity, and α the linear coefficient of thermal expansion (3α = volume coefficient of expansion). The correction is small for metals, but demonstrates that elasticity and thermal expansion are involved in the vibration of atoms. For most materials, the Debye characteristic temperature is below room temperature, but the lighter atoms such as beryllium, boron, and carbon have high values of elastic modulus and Θ_D is above room temperature. Figure 5.2 shows the variation of specific heat per gramme-atom with atomic number, and the general hyperbola on which all these results obeying the Dulong and Petit law fall. The exceptions are those that have a high Debye temperature.

5.3. *Thermal expansion*

If the atoms vibrated about an equilibrium position with perfect symmetry, there would be no thermal expansion. The forces of attraction and repulsion do not change with temperature, but they vary independently with distance between atoms, so that the atoms must vibrate asymmetrically.

The symbol ε_1 is used to denote the incoming semi-amplitude of vibration of the atoms. As the atoms come closer together, the force of repulsion increases, and so they move apart again. Let the outgoing semi-amplitude of vibration be ε_2. This time, the forces of repulsion decrease, and are overcome by attraction so that the atoms swing together again. ε_2 must therefore exceed ε_1 for a resultant elongation of the lattice, represented by ε. Figure 2.2 illustrates this effect of increased amplitude on the potential energy curve and the fact that ε_2 must exceed ε_1 owing to the balance between attraction and repulsion. From Figure 2.2, $\varepsilon_1 = r_0 - r_1$ and $\varepsilon_2 = r_2 - r_0$, where r_0 is the mean position, at which point dE/dr is zero. The mean semi-amplitude of vibration may be written $\bar{\varepsilon}$, and the elongation ε (thermal expansion) is then given by a normal sinusoidal vibration equation

$$\varepsilon = \bar{\varepsilon} \cos \omega t$$

where ω is the velocity of vibration and t the time. The thermal energy of the vibration will be proportional to $\bar{\varepsilon}^2$. The equation of the curve in Figure 2.2 is that given by MIE,

$$\phi(r) = -\frac{a}{r^m} + \frac{b}{r^n}$$

From this, the value of $\varepsilon_1 + \varepsilon_2$ can be shown to be

$$\varepsilon_1 + \varepsilon_2 = \tfrac{1}{3}(m + n + 3)\bar{\varepsilon}^2$$

and the average value of ε, the elongation, as

$$\varepsilon_{av} = \tfrac{1}{4}(m + n + 3)\bar{\varepsilon}^2 \tag{5.8}$$

Now the coefficient of thermal expansion α is defined as $d\varepsilon_{av}/dT$. From equation (5.8), ε_{av} is proportional to $\bar{\varepsilon}^2$ and to thermal energy; hence $d\varepsilon_{av}dT(= \alpha)$ must be proportional to $dE/dT = C_v$. Thus, the thermal coefficient of expansion is related to specific heat, and both

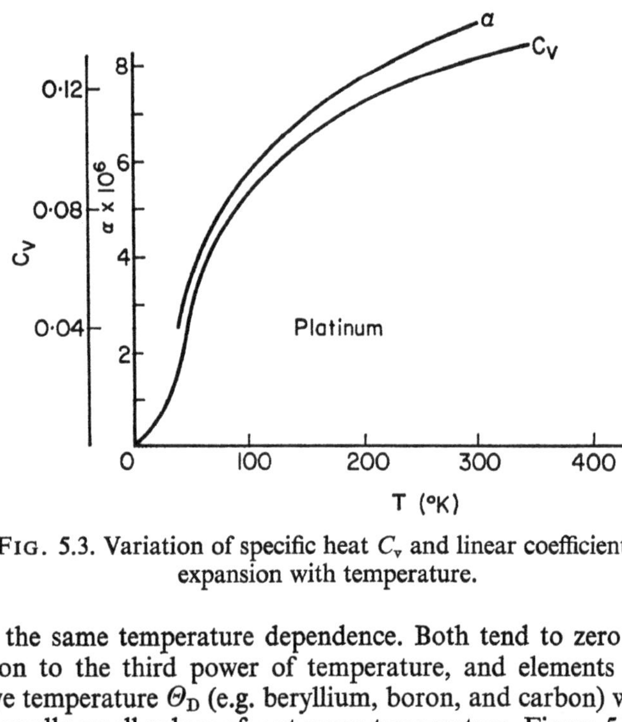

FIG. 5.3. Variation of specific heat C_v and linear coefficient of expansion with temperature.

have the same temperature dependence. Both tend to zero in proportion to the third power of temperature, and elements of high Debye temperature Θ_D (e.g. beryllium, boron, and carbon) will have abnormally small values of α at room temperature. Figure 5.3 shows the values of α and C_v for platinum between absolute zero and the boiling point of water.

When a body is compressed or expanded by mechanical forces, the potential energy ϕ is given by

$$\phi = \frac{K(\Delta V^2)}{2V}$$

where V is the volume, ΔV the change in volume, and K the bulk modulus of elasticity. Since $\Delta V/V = 3\varepsilon$, then $\phi = \tfrac{1}{2}KV(3\varepsilon)^2$. The thermal energy E may be considered entirely potential energy at the point of greatest deviation, where $\varepsilon = \bar{\varepsilon}$; thus $E_T = \frac{1}{2}KV(3\bar{\varepsilon})^2$. Substituting in equation (5.8),

$$\varepsilon_{av} = \frac{m+n+3}{4}\frac{2}{9KV}E_T$$

and differentiating with respect to temperature again, and remembering $\varepsilon_{av}/dT = \alpha$ and $dE_T/dT = C_v$

$$\alpha = \frac{m+n+3}{4}\frac{2}{9KV}C_v$$

or

$$\frac{KV3\alpha}{C_v} = \frac{m+n+3}{6} = \gamma \qquad (5.9)$$

where γ is called *Grüneisen's number.*

The value of γ varies between narrow limits. By substituting for the values of m and n associated with different types of lattice, we can obtain a correlation between K, α, and C_v.

For ionic lattices, $m = 1$ and $n = 2$, so that $\gamma = 1$. For bonds involving Van der Waals' forces, homopolar bonds in gas molecules, etc., $m = 6$ and $n = 9$, and $\gamma = 3$. Most materials fall within these limits. Measurement of K, V, 3_v, and C_v enables an assessment to be made of the constants m and n of the Mie equation. In body-centred cubic structures, the average value of γ is 1·6, and so $m + n = 7$. The assumption is that $m = 2$ and $n = 5$. In face-centred cubic metals, the average value of γ is 2·1, and $m + n = 10$, with the assumption that $m = 1$ and $n = 9$, taking other evidence into consideration.

B.C.C. metals:
γFe = 1·6; γMo = 1·57; γW = 1·62; γNa = 1·25

F.C.C. metals:
γAl = 2·17; γCu = 1·96; γAg = 2·40; γPb = 2·73

It would appear that sodium ($\gamma = 1·25$) has a lattice of more ionic nature ($\gamma = 1$) (see page 30) whereas lead ($\gamma = 2·73$) more closely resembles the homopolar-bond type ($\gamma = 3$) (see page 33). As γ is a constant for the lattice, $KV3\alpha/C_v$ is constant. The ratio V/C_v will not vary much between elements (Section 5.2), so that values of α and $1/K$ will vary in a similar manner when plotted against atomic number Z. The implication is that α is largely controlled by the valency of the ions.

In the short periods of the periodic system, the strength of bonding increases from groups I to IV, since the electrons involved (valency electrons) are in hybrid (*sp*) states (see page 26). The proportion of electrons in *p* states increases with group number, so that the strength increases and α decreases, as shown in Figure 5.4.

In the first long period, where 3*d* electrons are also involved, there are hybrid states between (4*s*)(3*d*) or between (4*s*)(3*d*)(4*p*), since these separate states are so very close in energy level (see page 15). The same general principle of bonding will therefore apply in passing from potassium to vanadium. From vanadium to chromium, the

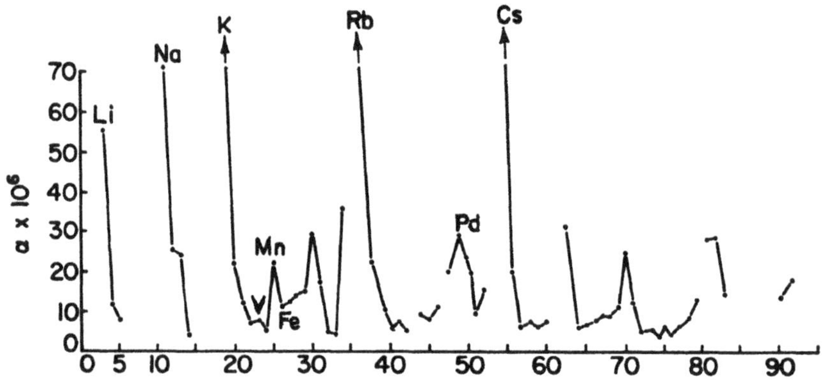

Atomic number Z

FIG. 5.4. Variation of linear coefficient of expansion with atomic number.

strength increase is slightly reduced, actually decreasing with manganese, and then increasing with iron, cobalt, and nickel. These variations, while following the general principles outlined, are obviously affected by the more complex electronic structure.

In the other long periods there is a more steady increase in binding through to group VIA, where (5*s*)(4*d*) or (5*s*)(4*d*)(5*p*) electrons may be hybridized. There is a decrease at groups VIIIB and VIIIC (ruthenium to palladium) and another loss of strength in passing to group IB with silver. Some further change obviously occurs in the hybridization of electrons or their influence on *K* and α. (This may be associated with antiparallel spin.) The third long period is even more complicated by the rare earths and the introduction of 4*f* electrons.

5.4. *Melting*

Figure 2.2, the graphical form of the Mie equation for the forces of attraction and repulsion, shows that the resultant curve for potential energy has a point of inflexion, at which the forces of attraction begin to approach their maximum value. When the temperature of a body is raised so that the potential energy approaches this point, the outgoing semi-amplitude of vibration of the atoms begins to exceed the ingoing semi-amplitude at an ever-increasing rate. The crystal lattice is then on the point of break-up, for the attraction between any two atoms may be overcome, and the atoms remain apart. A little later so many bonds are broken that only portions of the original lattice remain coherent, and the normal long-range order of the crystal gives way to a much shorter-range order. When the break-up reaches a rapid rate, the thermal energy input is utilized in completing the break-up of enough of the atomic bonds, and in a pure element a momentary pause in the rise of temperature occurs. This is the melting point, and the thermal energy consumed is the latent heat of fusion.

At the point of inflexion (Figure 2.2) it can be shown that

$$r_i = \frac{m+n+2}{m+n+1} r_0$$

Since $r_i = r_0(1 + \varepsilon_i)$ (Section 5.3)

then
$$\varepsilon_i = \frac{1}{m+n+1}$$

where ε_i is the outgoing amplitude at the melting point.

The average amplitude ε_{av} is given by

$$\varepsilon_{av} = \tfrac{1}{4}(m+n+3)\bar{\varepsilon}^2 \qquad \text{(5.8, page 110)}$$

Thus, substituting ε_i for $\bar{\varepsilon}$

$$\varepsilon_{m.p.} = \frac{1}{4}\frac{m+n+3}{(m+n+1)^2} \qquad (5.10)$$

where $\varepsilon_{m.p.}$ is the linear thermal expansion of the body at the melting point.

The volume expansion $\Delta V/V$ is approximately $3\varepsilon_{m.p.}$; hence,

$$\Delta V/V_{m.p.} = \frac{3}{4}\frac{(m+n+3)}{(m+n+1)^2}$$

Substituting Gruneisen's number $\gamma = \tfrac{1}{6}(m+n+3)$,

$$\Delta V/V \text{ at melting} = 9\gamma/2(6\gamma - 2)^2$$
$$\simeq 1/8\gamma$$

The average value of γ for many substances is 2, so that the volume expansion at melting will be approximately 6%. For those elements with γ nearer to 1, the volume expansion will be more nearly 10%. This reasoning indicates that all substances expand by roughly the same amount from absolute zero to the melting point, provided that the value of γ is similar. Metals of high melting point such as tungsten are difficult to contain at the melting point, so that very accurate measurements of the expansion on melting are virtually impossible. By extrapolating thermal expansion measurements on the solid and using a value for Gruneisen's number based on other data a reason-

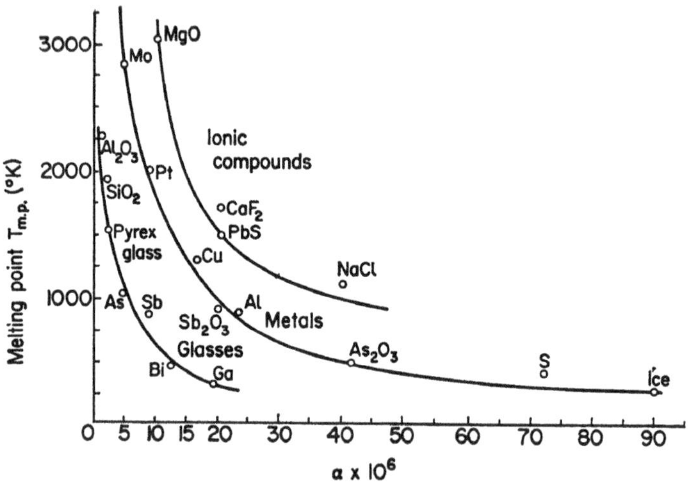

FIG. 5.5. Relationship between melting point and linear coefficient of expansion.

able estimate of the expansion on melting, or the contraction on solidification, can be made. Similar predictions can be made for other substances, if γ can be estimated. Plotting the melting point $T_{m.p.}$ against α, the linear coefficient of expansion, the curve is hyperbolic, since $3\alpha T_{m.p.}$ is a constant. However, the value of γ varies, and there will be separate curves corresponding to classes of substance for which γ is different. Many ionic compounds, e.g. alkali metal halides and some oxides, frequently show a low value of γ ($m = 1$, $n = 2$) and therefore a high value of $\Delta V/V$. The metals and remaining ionic compounds mostly fall near the average value of $\gamma = 2$, and thus $3\alpha T_{m.p.}$ is approximately 6%. The majority of amorphous structures, and those crystals showing homopolar-type binding (including metals such as arsenic, antimony, and bismuth), tend to a high γ ($m + n > 10$), so that $3\alpha T_{\text{m.p.}}$ is low, approximately 2 to 4%. Figure 5.5 shows the three curves obtained.

These curves can be used to demonstrate the feasibility of joining or coating one substance with another by fusion or softening. Metals to be sealed into glass or silica must have coefficients of expansion similar to that of the glass in order to prevent strains through differential contraction. The glass must soften at a temperature below the melting point of the metal. Since glasses fall on the lower curve of Figure 5.5, then, for the same value of α, they soften at a lower temperature than the metals on the upper curve. It is not possible, however, to make seals with arsenic, antimony, or bismuth, since they fall on the same curve as the glasses, nor would it be possible to seal aluminium, say, in calcium fluoride. The latter is on the upper curve. The vitreous enamels, used for domestic and industrial protection of steel or iron, obey the necessary relationship as for glass, and consequently can be melted on to ferrous metals for coating. They do not crack or spall on cooling, provided the composition is adjusted to give a similar value of α.

Equation (5.9), Grüneisen's rule, indicated that

$$\frac{KV3\alpha}{C_v} = \gamma$$

and since $\qquad 3\alpha T_{\text{m.p.}} \simeq \Delta V/V \quad \dfrac{KV}{C_v}\dfrac{\Delta V}{VT_{\text{m.p.}}} \simeq \gamma$

but $\Delta V/V \simeq 1/8\gamma$ (see above).

Thus $\qquad \dfrac{KV}{C_v}\dfrac{1}{8\gamma T_{\text{m.p.}}} \simeq \gamma \quad \text{or} \quad T_{\text{m.p.}} \simeq KV/C_v 8\gamma^2$

For metals $\gamma \simeq 2$, $\qquad T_{\text{m.p.}} \simeq KV/32C_v$

C_v is approximately $3R = 25$ joules gramme-atom^{-1} degK^{-1} (Dulong and Petit) at high temperatures, so that

$$T_{\text{m.p.}} \simeq KV/32 \times 25 = 12{\cdot}5 \times 10^{-4}\, KV$$

This gives an indication of the cohesion of metals in terms of melting point.

5.5. *Anisotropy*

The previous argument assumed that the vibration of atoms was uniform in three mutually perpendicular directions. Where the crystal lattice has unequal axes or angles, this cannot be true, and there will be three maximum frequencies and three values of the

Debye characteristic temperature. The specific heat of such a substance will be the sum of three similar functions in which Θ_D varies.

As an example, a hexagonal lattice such as zinc can be defined by two such values, since the three hexagonal axes *a* are equal in length and in angular spacing, the fourth axis *c* being different in length and at 90° to the other three. Thus, along the *a* axes, i.e. in the basal plane, there is a closer packing of atoms, a high elastic modulus due to the stronger binding, and a high Debye characteristic temperature,

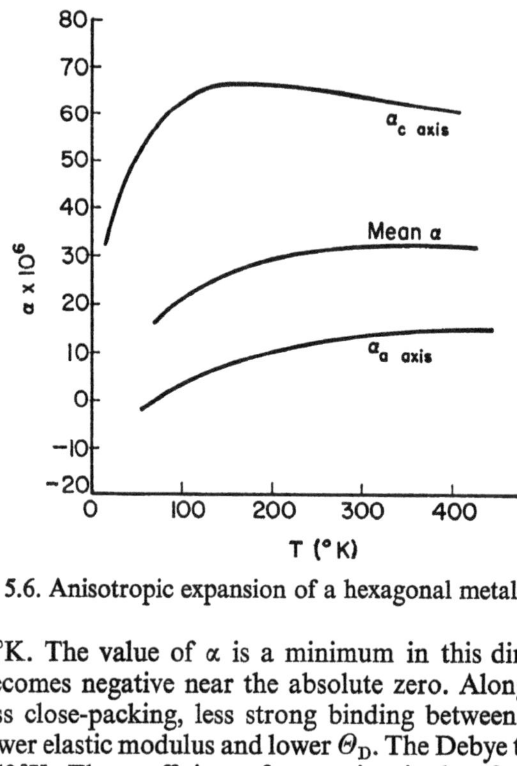

FIG. 5.6. Anisotropic expansion of a hexagonal metal—zinc.

$\Theta_D = 320$°K. The value of α is a minimum in this direction, and actually becomes negative near the absolute zero. Along the *c* axis there is less close-packing, less strong binding between the atoms, and so a lower elastic modulus and lower Θ_D. The Debye temperature is in fact 50°K. The coefficient of expansion is therefore high, but decreases at higher temperatures and falls again to zero as temperature approaches absolute zero. Figure 5.6 shows the variations of the two values of α with temperature, and the computed overall expansion of polycrystalline zinc which results. Chapter 7 discusses some consequences of this anisotropy.

5.6. *Conductivity*

5.6.1. *Thermal conductivity of insulators*

The application of thermal energy to one end of a cylindrical body will cause the atoms to vibrate as explained above. The transmission of energy along the cylinder will occur, in an attempt to decrease the free energy of the heated end, even if the source of heat is removed. If the body is an insulator, the electrons in the higher energy levels cannot be excited into conduction bonds, and therefore do not contribute to the transmission of energy. Only the lattice vibrations will do so. The factors affecting the transmission will be the inertia of the atoms (density), the effect of thermal energy on the vibration characteristics (specific heat), and the distance apart of the atoms (lattice constants). The ease of transmission will be the thermal conductivity, and can be defined by the expression

$$k_t = \rho c_p l s \tag{5.11}$$

where ρ is the density, c_p the specific heat at constant pressure, l the mean free path of the thermal wave, and s the velocity of sound. The heat capacity per mole C_p is equal to $\rho c_p . V$, where V is the volume, but $V = N_0 d^3$ where N_0 is Avogadro's number and d is the lattice parameter.

Hence $$C_p = \rho c_p N_0 d^3$$

C_p is approximately equal to $3R = 3N_0 k$, where k is Boltzmann's constant.

Therefore $$\rho c_p = \frac{3N_0 k}{N_0 d^3} = \frac{3k}{d^3}$$

Substituting in the equation for thermal conductivity (5.11), and assuming $l \simeq d$,

$$k_t \simeq 3ks/d^2 \tag{5.12}$$

The thermal conductivity of an insulator should then be inversely proportional to the square of the lattice parameter and directly proportional to the velocity of sound in the material.

The mean free path l depends on the thermal agitation of the lattice, and must be affected by any defects in the structure, such as dislocations and vacant lattice sites, and by impurity atoms. The thermal agitation is obviously temperature-dependent; the defects and impurities are not. The thermal conductivity of the lattice decreases in almost inverse proportion to the thermal energy, as the mean free path decreases. In less regular lattices, such as those of amorphous materials, there is a reduced probability of attenuation of the thermal wave, by virtue of the variable distances between the atoms, and the thermal conductivity is proportional to the thermal energy content.

Substances that can exist in crystalline or amorphous forms show very different curves for k_t against temperature (Figure 5.7). Where a heterogeneous substance contains some crystalline substances in an amorphous matrix, then the resultant conductivity will vary according to the substance that exerts the maximum effect. This is particularly important with refractory bricks, particularly those made from silica.

Materials of variable thermal conductivity may also vary in heat

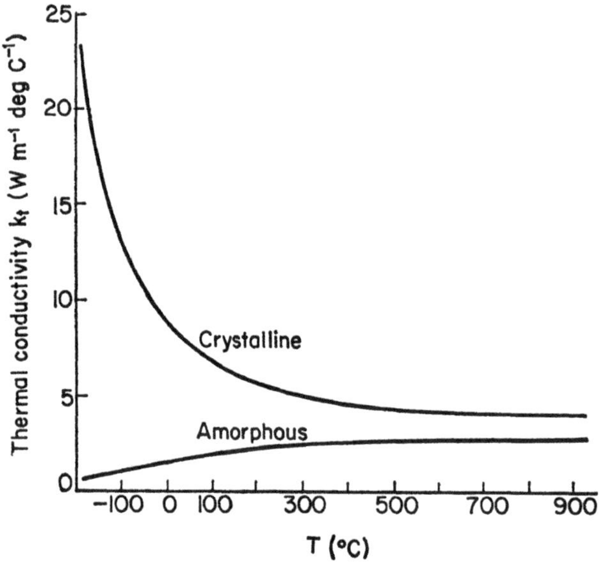

FIG. 5.7. Variation of thermal conductivity between crystalline and amorphous forms of SiO_2.

capacity. The rate at which the temperature rises in them is important in applications of thermal insulation to buildings, plant, and persons. A good insulator does not necessarily remain cold but if its heat capacity is low, the material is soon saturated and does not go on absorbing heat from the source. The parameter used to describe this is the *thermal diffusivity* a, the ratio of thermal conductivity to heat capacity,

$$a = \frac{k_t}{\rho c_p}$$

Metals have a high value of a, because their conductivity k_t is so high. Insulating materials are frequently porous. This gives a low value for both k_t and ρc_p, although a may not be any smaller than that for denser material. The propagation of temperature is therefore controlled by both k_t and ρc_p.

A similar parameter can be defined to indicate the rate of abstraction of heat when a cold and a hot surface are placed in contact. The contact coefficient b is defined by the square root of the product of conductivity and heat capacity:

$$b = \sqrt{(k_t \rho c_p)}$$

When b is small the temperature difference between the interior and the surface of a body where heat is applied or removed is high. Good insulating materials have low values of b, since both k_t and ρc_p are low. Such materials are applied to floor coverings, upholstery, and handles for this reason. Cotton sheets feel cold to the body in winter, owing to the higher value of b, which is due to a higher value of k_t as compared to wool. Metals feel cold to the touch, largely as a result of high k_t also. Textiles take up water during the night in unheated rooms. They have low b values, which causes them to cool off more rapidly than other materials. Water thus condenses on textiles, as it does on wooden bridges on a summer evening, whereas stone remains dry. Next day, textiles will heat up rapidly and give out their water. If the outside temperature is low, this water will condense on the windows or metals first. The latter have high b values and thus heat up more slowly. The condensation of water on glasses or metal objects brought from a cold to a heated room is another example.

The value of b for metals is remarkably independent of temperature, although k_t and ρc_p are both temperature-dependent for metals. This fact facilitates some of the heat-transfer calculations applied to metals.

5.6.2. *Thermal shock*

A sudden impulse of thermal energy can impose severe strains on materials, owing to the differential expansion between the interior and surface layers. The cracking of thick glass when placed in boiling water is a well-known example. Attempts to define a suitable parameter have varied according to the assessment of importance of thermal and mechanical properties.

The original definition is

$$sp_1 = \frac{a^{1/2}\sigma}{E\alpha}$$

where sp is the shock parameter, a the thermal diffusivity, σ the tensile strength, E Young's modulus, and α the coefficient of linear thermal expansion.

The derivation of the original parameter visualizes a localized temperature gradient created by the input of thermal energy, which is directly proportional to specific heat and inversely to conductivity, and hence proportional to $a^{-1/2}$. The strain produced results from differential thermal expansion, and is proportional to $\alpha/\sqrt{a}$. The internal stress is related to this strain by the modulus of elasticity, and hence is proportional to $E\alpha/\sqrt{a}$. The actual stress must then be related to tensile strength by the constant sp_1;

$$\frac{sp_1 E\alpha}{\sqrt{a}} = \sigma$$

If the impulse of energy is extremely localized, on the surface (as when a beam of radiation impinges on it), then the contact coefficient b is involved, and we can substitute b for a.

$$sp_2 = \frac{b^{1/2}\sigma}{E\alpha}$$

Where the whole body is immersed in a heated enclosure, the temperature gradient is no longer localized, and only conductivity is involved, because the heat capacity of the body is negligible compared with that of the enclosure; hence

$$sp_3 = k_t\sigma/E\alpha$$

Poisson's ratio ν, the ratio of longitudinal to transverse strain in tension, has been suggested as the important variable, as in

$$sp_4 = (1 - \nu)\sigma/E\alpha$$

For conditions in which the thermal conductivity would appear paramount, the third parameter $sp_3 = k_t\sigma/E\alpha$ may be developed. For substance with homopolar bonding, the material very often behaves in a brittle fashion, since the bonds can be separated before much deformation has taken place. (These materials are not necessarily weak.) Here $\gamma > 2$ $(m + n > 10)$, and since $\gamma = K3\alpha V/C_p$ (equation (5.9)) we can write E for K, and substitute $E\alpha \simeq C_p/V \simeq \rho c_p$.

Then $$sp_3 = k_t\sigma/\rho c_p = a\sigma$$

The value of a does not vary much between brittle materials, and hence tensile strength would appear to be a reasonable guide to thermal shock for this class of materials, in which $\gamma \simeq 3$.

Amongst the materials showing high resistance to thermal shock are oxides of magnesium, beryllium, and silicon, the complex oxides such as $FeO.Cr_2O_3$ and $MgO.Cr_2O_3$, and many metal carbides. Most of these materials possess only moderate thermal conductivity, but have low specific heat, low coefficient of thermal expansion, and high tensile strength. Any of the above parameters therefore indicates a high resistance to thermal shock. Carbides have high thermal conductivity, as do the metals, so that sp_3 (above) indicates resistance to shock in both cases. The metals are limited by a decrease in strength at high temperatures, whereas oxides and carbides have high melting points and so do not lose strength so rapidly with rise in temperature. Since they lack plasticity they tend to crack, whereas metals merely distort.

5.6.3. *Thermal conductivity of good conductors*

The thermal conductivity of insulators has been discussed in terms of the lattice vibrations only. The electrons play no part in the process. In semiconductors, however, there is the effect of temperature or impurities on the population of electron energy levels to take into account. The mean free path of the electrons in metals (and of the electrons and holes in semiconductors; see page 47) affects the value of conductivity and its temperature coefficient. In an ideal lattice there is no resistance to the electrons mobility in the lattice, and no transfer of energy by collision. Lattices are not ideal, however, and atoms are not at rest, so that there is a probability of collision, and therefore of a transfer of momentum. The mean free path of electrons is thus partly dependent on the thermal energy of the lattice. The regularity of the lattice is also affected by impurity atoms and by lattice defects, but these are not temperature-dependent. The electron contribution to conductivity therefore consists of two terms, one dependent on thermal energy and the other on structure only.

Metals in general have both high thermal conductivity and high electrical conductivity. It would be expected, therefore, since both parameters depend partly on the lattice and partly on the electrons, that they may be related. The expression

$$k_t/k_{el}T = \text{constant}$$

where k_t is the thermal conductivity, k_{el} the electrical conductivity, and T the absolute temperature, is known as the *Wiedemann–Franz–Lorentz* relationship, and the Lorentz constant has the theoretical value $2{\cdot}45 \times 10^{-8}$ V^2 degK^{-2}. Normally, the electrical resistivity ρ_e is used, rather than conductivity:

$$k_t\rho_e/T = \text{constant}$$

This allows measurement of electrical resistivity to be made as a guide to thermal conductivity when the latter may be difficult to measure, e.g. on small samples of rare materials.

The electron influence on the thermal conductivity of metals is normally far greater than the lattice contribution. When this is the case, the Lorentz constant holds reasonably well. In semiconductors the number of electrons free enough to wander through the lattice is small, and hence their effect on conductivity decreases and the Wiedemann–Franz relation does not hold.

At temperatures approaching absolute zero, the electron-scattering contribution to conductivity is proportional to T^{-2}, whereas the defect contribution is proportional to T.

Thus, $$k_e = aT^{-2} + bT$$

where k_e is the electron contribution, and

$$k_t = k_e + k_{vib}$$

where k_{vib} is the lattice-vibration contribution. Thermal conductivity therefore shows a maximum value near absolute zero, this maximum occurring nearer to absolute zero the more pure and more structurally perfect the metal.

At higher temperatures, the electron-scattering effect is directly proportional to temperature, and the mean free path is therefore proportional to T^{-1}. The energy of the electrons is increased proportionally to temperature, however, and the product of these two effects becomes almost independent of temperature. The value k_t thus becomes more independent of temperature at much higher temperatures.

The lattice contribution k_{vib} depends on a term proportional to T^2 at low temperatures, owing to the scattering of the atom vibrations by the electron movements. Lattice defects due to dislocations have a similar effect. Impurity atoms are responsible for a term dependent directly on T, and so the measurement of thermal conductivity at low temperatures can assist in separation of the effects due to dislocations from those due to impurities. This separation cannot easily be made from electrical resistivity measurements. Since the value of k_{vib} is much larger at low temperatures than it is at higher ones (by nearly 50%), the measurements of k_t at low temperatures are reasonably sensitive.

6

Electrical, Magnetic, and Optical Properties

6.1. *Electrical conductivity*

Electrical conductivity has already been referred to in discussing the bonding of atoms. It is largely determined by the number of electrons in the conduction band, and by the number of *holes* in the valence band. The mean free path of the electrons in metals (and that of electrons and holes in semiconductors) will affect the value of electrical conductivity and its temperature coefficient. There is some contribution to electrical conductivity by the lattice also, as in thermal conductivity, but this contribution is now much the smaller of the two.

As there is an obvious correlation between electrical conductivity and the movements of electrons and atoms in the lattice, both movements being affected by temperature changes, it is convenient to refer the values of specific electrical conductivity to similar conditions of atomic vibrations, as was the case for thermal conductivity. This can be done by comparing the conductivity with respect to atoms per unit volume at a given fraction of the Debye characteristic temperature. An 'order of merit' prepared in this way shows the highest electrical conductivities to be characteristic of copper, silver, gold, and the alkali metals. These are normally univalent metals. As the number of valency electrons per atom increases, there can be less scope for electrons to move into higher levels, owing to restrictions of quantum-mechanical principles and the *zone* theories of electron distribution. Consideration of these more complex ideas is beyond the scope of this book.

Comparison of the conductivities of the alkaline earth metals (group II) with those of group I shows a sharp drop in the values (Na→Mg; K→Ca). On passing to the trivalent elements (Mg→Al; Ca→Sc) the effect is variable. Aluminium is a very good conductor. All elements with *p* electrons, like aluminium, do not have good electrical conductivity, but in fact comprise the bulk of semiconductors (silicon, germanium, etc.). There is, however, commonly a rise in conductivity with increase in atomic number in any one group containing *p* electrons, as exemplified by the rise from arsenic to

TABLE V

Electrical resistivity of some natural elements, alloys, and compounds
(Ω m units at 20°C unless otherwise stated)

Element		Resistivity (Ω m)	Element		Resistivity (Ω m)
Silver	Ag	$1{\cdot}6 \times 10^{-8}$	Niobium	Nb	$1{\cdot}46 \times 10^{-7}$
Copper	Cu	$1{\cdot}67 \times 10^{-8}$	Vanadium	V	$1{\cdot}95 \times 10^{-7}$
Gold	Au	$2{\cdot}44 \times 10^{-8}$	Caesium	Cs	$2{\cdot}1 \times 10^{-7}$
Aluminium	Al	$2{\cdot}66 \times 10^{-8}$	Lead	Pb	$2{\cdot}1 \times 10^{-7}$
Beryllium	Be	$3{\cdot}8 \times 10^{-8}$	Uranium	U	$2{\cdot}9 \times 10^{-7}$
Magnesium	Mg	$3{\cdot}9 \times 10^{-8}$	Arsenic	As	$3{\cdot}5 \times 10^{-7}$
Calcium	Ca	$4{\cdot}1 \times 10^{-8}$	Antimony	Sb	$4{\cdot}2 \times 10^{-7}$
Sodium	Na	$4{\cdot}3 \times 10^{-8}$	Titanium	Ti	$4{\cdot}21 \times 10^{-7}$
		(0°C)	Zirconium	Zr	$4{\cdot}46 \times 10^{-7}$
Rhodium	Rh	$4{\cdot}7 \times 10^{-8}$	Barium	Ba	$5{\cdot}0 \times 10^{-7}$
Iridium	Ir	$5{\cdot}3 \times 10^{-8}$	Mercury	Hg	$9{\cdot}6 \times 10^{-7}$
Tungsten	W	$5{\cdot}45 \times 10^{-8}$	Bismuth	Bi	$1{\cdot}16 \times 10^{-6}$
Molybdenum	Mo	$5{\cdot}65 \times 10^{-8}$	Gadolinium	Gd	$1{\cdot}3 \times 10^{-6}$
Zinc	Zn	$5{\cdot}92 \times 10^{-8}$	Manganese	Mn	$2{\cdot}6 \times 10^{-6}$
Cobalt	Co	$6{\cdot}24 \times 10^{-8}$	Graphite	C	$1{\cdot}4 \times 10^{-5}$
Nickel	Ni	$6{\cdot}84 \times 10^{-8}$	Silicon	Si	$8{\cdot}5 \times 10^{-4}$
Cadmium	Cd	$7{\cdot}4 \times 10^{-8}$	Germanium	Ge	1×10^{-3}
Cerium	Ce	$7{\cdot}8 \times 10^{-8}$	Selenium	Se	8×10^{-2}
Iron	Fe	$8{\cdot}85 \times 10^{-8}$	Phosphorus	P	1×10^{-1}
Osmium	Os	$9{\cdot}1 \times 10^{-8}$	Chlorine	Cl	10 (−70°C)
Platinum	Pt	$1{\cdot}07 \times 10^{-7}$	Boron	B	$1{\cdot}8 \times 10^{4}$
Palladium	Pd	$1{\cdot}08 \times 10^{-7}$	Bromine	Br	$7{\cdot}8 \times 10^{10}$
Tin	Sn	$1{\cdot}15 \times 10^{-7}$	Diamond	C	5×10^{12} (15°C)
Chromium	Cr	$1{\cdot}35 \times 10^{-7}$	Sulphur	S	2×10^{15}
Tantalum	Ta	$1{\cdot}45 \times 10^{-7}$			

Alloy or compound	Resistivity (Ω m)	Alloy or compound	Resistivity (Ω m)
Duralumin	$3{\cdot}8 \times 10^{-8}$	Phenol formaldehyde resins	10^{12}
70/30 brass	$6{\cdot}2 \times 10^{-8}$	Polyethylene	10^{13}
90/10 bronze	$9{\cdot}5 \times 10^{-8}$	Nylon	10^{14}
Mild steel	$1{\cdot}7 \times 10^{-7}$	Quartz SiO_2 } Mica	9×10^{14}
70/30 monel	$4{\cdot}8 \times 10^{-7}$	Plate glass	10^{15}
18/8 stainless steel	$7{\cdot}0 \times 10^{-7}$	Pyrex	10^{16}
Silicon carbide SiC	$2{\cdot}5$ (at 1,100°C)	P.T.F.E.	2×10^{16}
Magnesium oxide MgO	10^5 (at 1,100°C)	Polystyrene	10^{18}
Fireclay	$1{\cdot}4 \times 10^8$	Fused silica	10^{20}

antimony to bismuth. The transition metals, containing *d* or *f* electrons, have relatively low electrical conductivity per atom, as with those metals crystallizing according to the 8 − *N* rule. Table V shows some typical values of electrical resistivity (the reciprocal of conductivity) for conductors, semiconductors, and insulators.

Reference was made in Chapter 5 (page 121) to the fact that electrons would not experience any resistance to their mobility through an ideal lattice, but that in the imperfect lattices of most solids there is a probability of collision and a transfer of momentum. The mean free path of an electron is then affected by a temperature-dependent probability, and a part of the electron contribution to electrical conductivity is temperature-dependent. Electrical conductivity is affected in a similar way to thermal conductivity by impurity atoms and defects in the lattice structure. The effect is independent of temperature.

At absolute zero, an ideal metal should have zero electrical resistance, but since structural imperfections do occur, a residual resistance remains. As the temperature is raised from absolute zero, the resistance changes proportionally with the fourth or fifth power of temperature, up to a level about one-tenth of the Debye characteristic temperature. Above this temperature, the contribution of thermal energy approaches its optimum, and the scattering effect on the electrons becomes directly proportional to absolute temperature at about $\frac{1}{2}\Theta_D$ to Θ_D. This relationship was first demonstrated by Mathiessen (see Figure 6.1).

$$\rho_e = T\frac{d\rho_0}{dT} - \rho_0 + \rho_{str} \tag{6.1}$$

where ρ_e is the electrical resistivity, ρ_0 the displacement from direct dependence on T, ρ_{str} the resistivity due to impurities and imperfections, and T the absolute temperature.

Figure 6.1 illustrates the importance of low-temperature resistivity measurements in the determination of structural imperfections. The value of ρ_{str} is increased by quenching, cold-working, or the presence of impurities. This simple measurement is a feasible test for purity.

When pressure is applied to a metal the electrical resistivity decreases, since the number of electrons per unit volume is increased.

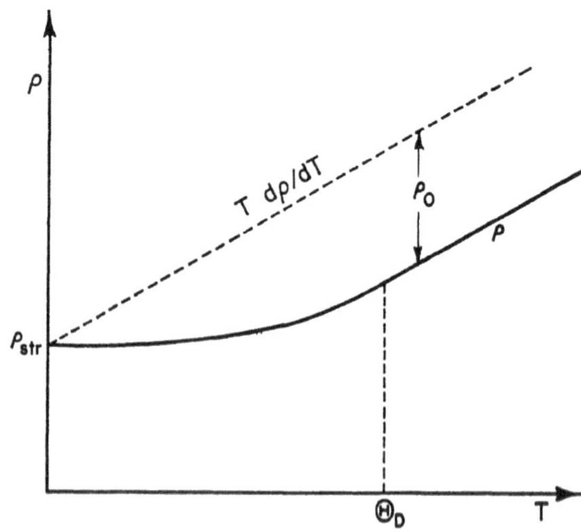

FIG. 6.1. Mathiessen's law.

This fact is utilized as the basis of electrical resistance strain gauges, whereby change in resistivity can be correlated with elongation.

The resistivity–temperature graphs of certain transition metals show a pronounced curvature at temperatures about Θ_D. This is believed to result from the hybridization of s and d electrons, and the effect on the electron-cloud–lattice perturbations. These effects disappear at higher temperatures. Near the melting point an abnormally large increase in resistivity occurs with some metals, which indicates that vacant lattice sites are being created in profusion as the long-range order of the lattice begins to change to short-range order. The atomic vibrations in the lattice become so large that quantum-mechanical considerations of the solid state no longer apply, and the resistivity rises.

A limited number of metals and compounds lose their resistivity before absolute zero is reached, and become *superconductors*. The exact theoretical explanation is still in doubt, but seems to be connected with the interaction between electrons and lattice vibrations. The transition temperature, at which the phenomenon occurs, varies; it is 0·4°K for titanium, 1·17°K for aluminium, and 9·2°K for niobium. For the compound Nb_4Sn the transition temperature is as high as 18·2°K. The crystal structure of the material markedly affects the superconductivity. White tin, with a tetragonal lattice, is superconducting below 3·72°K, whereas grey tin, with a cubic structure, is not a superconductor. Mercury has two transition temperatures, alpha Hg at 4·15°K, and beta Hg at 3·95°K. Bismuth becomes superconducting only when compressed, and beryllium only when deposited in thin films from the vapour phase. The superconducting state can be destroyed by the application of an external magnetic field. In the absence of any external magnetic field, an increase in the current passing through the conductor can itself create a sufficient change in magnetic field to restore the resistivity of the conductor.

6.2. *Ionic conductivity*

An ionic lattice (or ions in a solution) displays normal thermal agitation, and collisions between ions, other impurity atoms, or solvent molecules take place continuously. These movements are purely random in direction, and increase with rise of temperature. If, however, an external electric field is applied, e.g. if the sample is connected to a source of e.m.f., there will be a net movement of ions towards the poles of opposite sign. This movement is still affected by thermal agitation, and is therefore comparatively slow, by virtue of frequent collisions. The mobility of the ions increases with increasing temperature, and so the ionic conductivity will also rise. Since the ions are relatively large, their movement is much less, for a given field, than would be that of electrons, and ionic conductivity is therefore less than electronic conductivity.

In solid materials ionic conduction is strongly structure-sensitive, particularly with respect to impurities and flaws or defects in the lattice. The passage of current through an ionic conductor will gradually eliminate the impurities, by virtue of their movement to the poles. The conductivity then falls to the true value of the ionic crystal proper, and is limited only by ionic mobility, as referred to above. Diffusion is obviously paramount in this removal of impurities by electrolysis. The mobility of individual ions is obviously affected by their size, but the electronic charge must also play a part.

This explains why lithium can be more easily eliminated from quartz crystals than sodium, and why the magnesium ion, although smaller than sodium, is less mobile, owing to its divalent state.

6.3. *Semiconductors*

The limitations on the passage of current in metals and ionic lattices have been shown to depend largely on imperfections in the first case and on size factors in the second. In Chapter 2 the inherent difference between a conductor and a semiconductor was referred to. It was emphasized that in the semiconductor the *number* of electrons available for conduction purposes is important, and that this number varies with temperature and the magnitude of the electric field. At room temperature the resistivity values of semiconductors lie between those of conductors and insulators, differing from those of metals by a factor of roughly 10^6 (Table V).

At absolute zero the electrons in the structures typical of silicon and germanium are shared in homopolar bonds and are not available to carry current. As the temperature rises the electrons in the valency band, which are the bonding electrons, are increased in energy and, when the gap between the valency band and the conduction band is small enough, will subsequently move into the higher energy conduction band. These electrons are now less restricted in the conduction band, and can migrate through the crystal. The presence of impurities affects the electron movement by scattering, especially at low temperatures, and may in fact limit conduction by reducing the mobility of the electrons. In an intrinsic semiconductor, therefore, the characteristic decrease in resistivity with rise in temperature requires the material to be of a high standard of purity. The effect of certain types of impurity atom was also described in Chapter 2. Here, the requirement is either one more valency electron than the base atom or one less. The n-type semiconductor contains impurity atoms with one extra electron, which can be excited into the conduction band where the gap is too large for the base atom electrons to bridge the gap, or may increase the conductivity of an already intrinsic semiconductor. The p-type contains impurity atoms with one electron less than the base atom, but helps to bridge the gap by allowing electrons from the base atom to be excited from the valence band into the energy levels of the impurity atoms, thus allowing carriers in the valence band, now no longer full. With increasing temperature these holes in the valence band are now more free to migrate than they were (Figure 6.2).

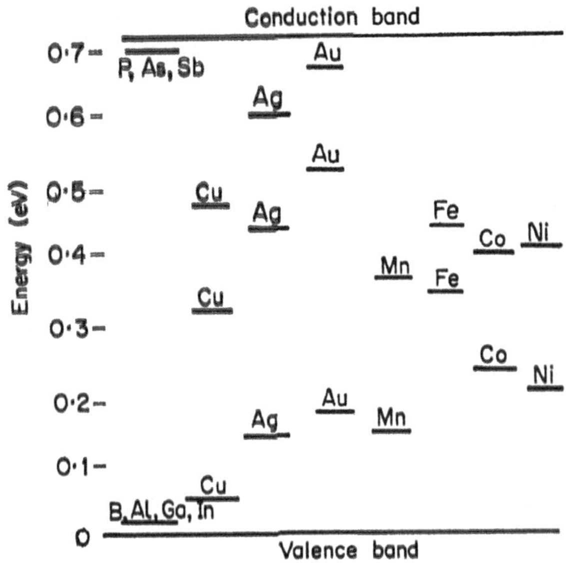

FIG. 6.2. Location of energy levels of impurities in germanium.

The two types of impurity semiconductors naturally show different characteristics, since the mechanism of their operation is different. Taken into conjunction with an already intrinsic semiconductor material such as silicon and germanium, they allow considerable flexibility in the design of semiconductor materials with particular characteristics of variation in resistivity with temperature and electric field. In pure germanium, for instance, the resistivity at room temperature is about 0·5 Ω m. The addition of arsenic (one extra electron—n-type) in a concentration of 10^{20} atoms/m^3, decreases this value to 0·2 Ω m. Further additions of arsenic, to ten times the previous concentration, produce a decrease in resistivity to 0·01 Ω m. Similarly, the same concentrations of gallium (one less electron—p-type) in germanium give comparable resistivity values, of 0·03 and 0·02 Ω m respectively.

The standard of purity required for intrinsic semiconductors, and the very close control required for impurity additions in the extrinsic type, has led to a considerable refinement of techniques in preparation of these materials. The technological problems of making contacts between conductor wires and the semiconductor material, and the use of different types of semiconductor material in combination, are also great. Many of these involve compatibility and alloying between dissimilar metals or metals and compounds. The scientific background to these difficulties is the same, however, as for less-refined materials, and is referred to in Chapter 8.

6.4. *Thermoelectric effects*

When dissimilar materials are in contact, and a current passes across the junction, heat may be generated, because electrons conduct thermal as well as electrical energy, and the proportions of each associated with different substances are not the same. Conversely, when heat is applied to such a junction, an e.m.f. may be generated. The first effect is known as the *Peltier effect* and the second as the *Seebeck effect*. The order of magnitude of the Peltier effect varies from normal metallic conductors ($\times 1$) up to values of $\times 100$ for metals whose structures obey the $8 - N$ rule, and $\times 1{,}000$ for semiconducting compounds.

A third effect, the *Thomson effect*, occurs when a conductor (or semiconductor) is carrying current and there is at the same time a temperature gradient along the conductor in the same direction. Heat is evolved as a result of the electron movement. The net result of these three effects is utilized in the thermocouple and the thermopile. (The latter consists of a number of thermocouples in series.) By choosing appropriate values of the Peltier and Thomson coefficients, two dissimilar materials can be arranged to produce a substantial e.m.f. if one junction is maintained at a high temperature and the other at a lower one (often taken at room temperature or that of melting ice). This couple can then, after suitable calibration, be used for temperature measurement.

6.5. *Magnetic properties*

In Chapter 1 the quantum-mechanical approach was used to describe the energy states of electrons. The motion of the electrons was described by the four quantum numbers, three of which involved the setting-up of magnetic fields as a result of angular momentum and electron spin. The physical picture of the *s* energy states was seen as a cloud of negative electricity, having spherical symmetry and no net angular momentum. Accordingly, no magnetic moment is associated with *s* levels from the standpoint of angular momentum. All other energy states have charge distributions that may be non-spherical. There is in these cases a net angular momentum and a consequent magnetic moment. Plate II illustrates this for $2p$ states in hydrogen. The magnetic moment was defined in terms of the *Bohr magneton* $eh/4\pi mc$ as a basic unit.

In the case of p states, a magnetic moment due to angular momentum can occur only if one or two of the quantum number m_l values are filled by electrons. If all three values ($m_l = +1, 0, -1$) are occupied, then the resultant magnetic moment is once more zero, since the three fully occupied p states again approach a spherical symmetry. There is therefore no resultant magnetic moment in the inert-gas structure, where all p states are occupied, but there will be magnetic moments in elements where the three p states are only partially occupied.

In the free atom any incompletely filled p, d, or f state will produce a resultant magnetic moment, owing to the angular momentum of the electrons. In the solid state this is affected by the influence of surrounding atoms in the lattice; all energy levels are now smeared into bands and may prevent the easy alignment of the magnetic dipole moments in the presence of a magnetic field. The orbital magnetic moment is said to be *quenched*, as happens in the iron group of elements. As the elements increase in atomic number, the f electrons occur deeper inside the atom, and are shielded from any effect of the surrounding lattice, so that the transition metals, with higher atomic number, retain their magnetic moments even in the solid state.

In addition to the orbital angular momentum producing a magnetic moment, mention was made in Chapter 1 of the magnetic effects produced by the electron spin. The pairing of electron spin affects this magnetic moment; the fewer electrons involved in antiparallel spin, the greater the moment may become. In the iron group, although the magnetic moment from angular-momentum effects is quenched in the solid state, there is a large moment produced by electron spin.

The structure of the atomic nucleus hardly concerns this book, but mention was made at the end of Chapter 1 of the neutron/proton ratio. The fine structure of the nucleus is in fact not dissimilar to that of the electron states, and spin of protons produces a further nuclear magnetic moment. Since the value of this nuclear magnetic moment is three orders of magnitude smaller than that of the electrons, it will be neglected, bearing in mind that use can be made of the property to study molecular structure by nuclear magnetic resonance.

When a substance is placed in an external magnetic field, the various electronic factors that contribute to the magnetic moment of the atom produce an induced magnetic moment. *Lenz's law* states that an induced current produces a magnetic field to oppose the one applied externally, and this happens in the case of atoms, whose induced magnetic moments oppose the applied field. Bars of such substances tend to align themselves perpendicular to the applied field, and are referred to as *diamagnetic*. Removal of the field removes the induced opposing moment, and there are no after-effects. The ratio of intensity of magnetization to the applied field is called the *susceptibility*, χ, defined by the expression

$$\chi = I/H$$

where I is the induced magnetic field and H the applied magnetic field. For diamagnetics, χ is negative, and is largely governed by the outer electrons, whose angular momentum produces the greatest effect on the induced magnetic moment. The more closely spherical is the electron-cloud distribution, the more effective are the outer electrons. This is the case for inert gases and ions, and calculation of susceptibility based simply on the outer electron states produces close agreement with experiment. An increase in atomic number, which increases the equivalent outer radius of these effective outer electrons, also increases the diamagnetic susceptibility; e.g. Li^+ −0·7; Na^+ −6·1; and F^- −9·4; Cl^- −24·2 (all $\times 10^{-6}$). (Note: these are ions, not atoms in a solid crystal.) Anisotropic materials also show anisotropy in the diamagnetic susceptibility, but values of χ_D are small on the whole, antimony and graphite being among the highest at approximately 10^{-3}. Many superconductors are diamagnetic, and when an external field is applied to them the small induced electric currents that result from the induced magnetic field of the electrons are able to spread through the whole conductor in view of the negligible resistance offered to them. When the external field is removed, these effects may remain for long periods, whereas normal diamagnetics immediately lose the opposing induced field. This retention of the induced field enables superconductors to be used as *memory-storage* units in computers, the field being tapped off later.

As stated above, the angular momentum effects are weak, and the diamagnetism produced by them may be overshadowed by the stronger magnetic moments of electron spin. When there are complete electron shells, with pairs of antiparallel-spin electrons, the spin moments cancel each other out, and only diamagnetism remains. (This is why ions behave differently from the normal atom.) With unpaired electrons, the spin moments may be aligned parallel or antiparallel to any applied field. A lower free energy results from parallel alignment; the spin moments then tend to assist the applied field, and the susceptibility becomes positive. These substances are *paramagnetic*. CURIE first studied these substances systematically, and deduced the law for paramagnetic susceptibility:

$$\chi_p = C/T$$

where χ_p is the paramagnetic susceptibility, C the Curie constant, and T the absolute temperature.

The paramagnetic susceptibility is always positive, and normally so large that it swamps the diamagnetism due to orbital angular momentum and the nuclear spin effects. As stated previously, the inert gases, ions with complete outer electron shells, and elements crystallizing according to the $8 - N$ rule all tend to be weakly diamagnetic. The metals with s electrons (alkali and alkaline earth) tend to be weakly paramagnetic in the case of alkali metals (one s electron) and weakly diamagnetic with alkaline earth elements (two s electrons, antiparallel spin, closed shell). When the shells are completed by losing outer electrons (e.g. Na^+ ion) the ion is diamagnetic. Elements with one or two p electrons only (aluminium, gallium, indium) that do not crystallize according to the $8 - N$ rule tend to be weakly paramagnetic (not closed shells but parallel spin).

The transition elements, with d and f electrons, tend to be strongly paramagnetic, and those elements in the odd-numbered periodic groups to have the highest susceptibility.

Organic materials and the light elements (hydrogen, helium) are diamagnetic, usually with low values of susceptibility.

6.6. *Ferromagnetism*

The diamagnetism produced by the angular momentum of the electrons has already been shown to be quite weak, and easily surpassed by the paramagnetic influence of spin in unpaired electrons. When atoms containing unpaired spin electrons are bound together in the solid state, there is bound to be some interaction between them, provided that the unpaired electrons are not deep-seated and therefore shielded by the outermost electrons. This latter state exists in the rare earths, where the $4f$ electrons give rise to paramagnetism, but each atom is virtually screened from its neighbours. In the first long period, the initial appearance of $3d$ electrons is in the highest energy levels, and consequently interaction effects do take place. The greater the imbalance in electron spin, the greater the possible induction effects, and the group of elements iron, cobalt, nickel demonstrates this. Below a certain critical temperature these elements remain permanently magnetized after removal of the external field.

This critical temperature is called the *Curie point* or Curie temperature. Above it, the materials behave as normal paramagnetics, and the susceptibility is defined as

$$\chi_F = \frac{C}{T - \theta_c}$$

where T is the absolute temperature and θ_c the Curie temperature. Below the Curie temperature, the magnetic induction B does not increase proportionally to the applied field strength H, but tends to a saturation level. When the applied field is decreased, the magnetic induction decreases more slowly than it rose on magnetization, and retains a finite value when $H = 0$, called the *remanence* B_R. If the applied field is now reversed and slowly increased in this reverse direction, the value of B decreases finally to zero, and the value of applied field at this point is called the *coercive force* H_c. Further increase of H in this reverse direction magnetizes the material in a way similar to the previous saturation, and subsequent decrease of H through zero and reversal of direction again completes the familiar *hysteresis* loop or *B–H* curve (Figure 6.3).

Such substances are known as *ferromagnetics*. The *permeability* of

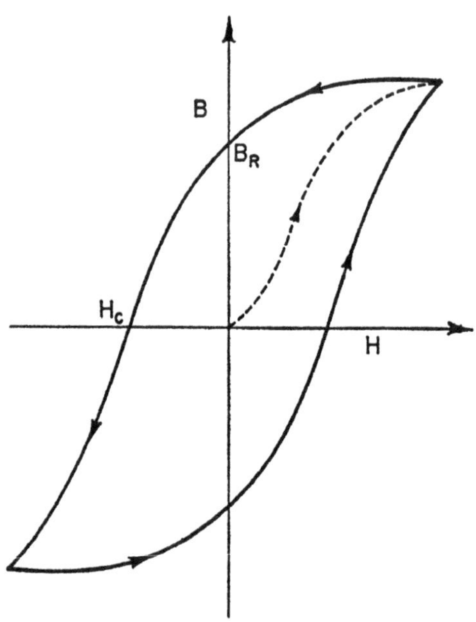

FIG. 6.3. The B–H magnetization curve.

the material is defined as the ratio of magnetic induction to applied field:

$$\mu = B/H$$

The discussion of electron configurations in the elements (in Chapter 1) emphasized the tendency for association of electrons in parallel spin rather than antiparallel, when a number of electrons were involved. In the case of the first long period transition elements, scandium to zinc, the $3d$ electrons build up to five in manganese, all of parallel spin. In iron, the first pairing results, with only one electron spin in one direction and five parallel in the opposite direction. Cobalt has seven $3d$ electrons, so that two are in one direction and five parallel in the opposite, making two pairs and three electrons unpaired. Nickel has three pairs and two unpaired. At copper, the

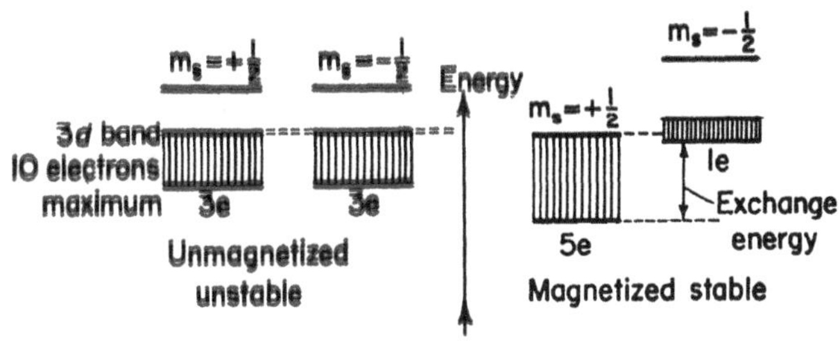

FIG. 6.4. $3d$ electron bands in iron.

situation changes with some hybridization, but even if this did not take place, there would be four pairs of electrons of antiparallel spin and only one electron unpaired. Zinc has all ten $3d$ electrons in five pairs. HEISENBERG first postulated that alignment of electrons in parallel spin took, place when atoms were brought close together, and SLATER and STONER use the band theory to reason that lower energy results from alignment of parallel spins, and this is therefore the more stable state. The width of the $3d$ band is important, however, for any increase in its width which added electrons of higher energy would cause greater instability. Figure 6.4 illustrates the $3d$ band in iron, with the electron spin directions separated into two distinct portions of the band, to show the lower energy obtained by alignment into parallel spin of five electrons. It is obvious that any further additions to the $m_s = +\frac{1}{2}$ band would mean higher energy. The separation of the lowest energy levels is called the *exchange energy*, and is the measure of the interaction effect discussed previously.

Since the width of the $3d$ band (after alignment of electrons into parallel spin) is bound to affect stability, the effect on the $3d$ band of bringing the atoms closer is also involved. At greater interatomic distances the exchange energy is decreased, but when the interatomic distance is very small there is a broadening of the $3d$ band. The condition that favours the parallel alignment of the electrons is that the ratio of one-half of the interatomic distance to the average radius of the $3d$ shell is greater than 1·5. This ratio for iron is 1·63, for cobalt 1·82, and for nickel 1·98. Figure 6·5 shows the distribution for these three metals and for the two preceding ones in the period.

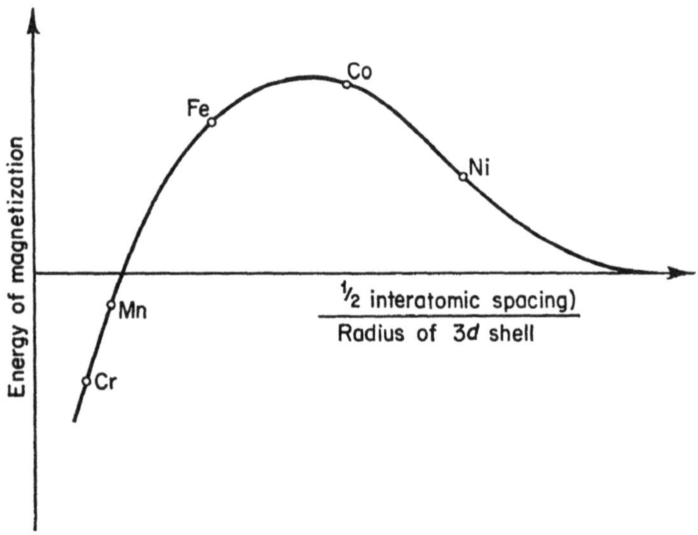

FIG. 6.5. Variations in equivalent exchange energy with spacing for some transition metals of the first long period.

Elements such as manganese are so close to the required condition that the addition of other elements to manganese in compounds or alloys may affect the interatomic distance favourably, and give rise to ferromagnetism. The addition of aluminium and copper can do this, as exemplified by the Heusler range of alloys. The compound of manganese with nitrogen is also ferromagnetic, for the same reason. Manganese alone is strongly paramagnetic.

The theoretical aspects discussed above do not explain the shape of the B–H curve, or the fact that a polycrystal of iron is not always magnetized. WEISS first postulated the idea that separate areas exist in a ferromagnetic material, called *domains*, in which the alignment of unpaired electron spin is parallel within any one domain, but varies from one domain to the next through all possible orientations. An overall zero magnetic moment results when all orientations of spin are present. When an external field is applied some domain boundaries are displaced, such that domains whose spin orientation lies parallel to the applied field tend to grow at the expense of those in other orientations. As the field strength increases, so the extent of domain boundary displacement increases. Approaching saturation, the domain growth and rotation of spin direction in

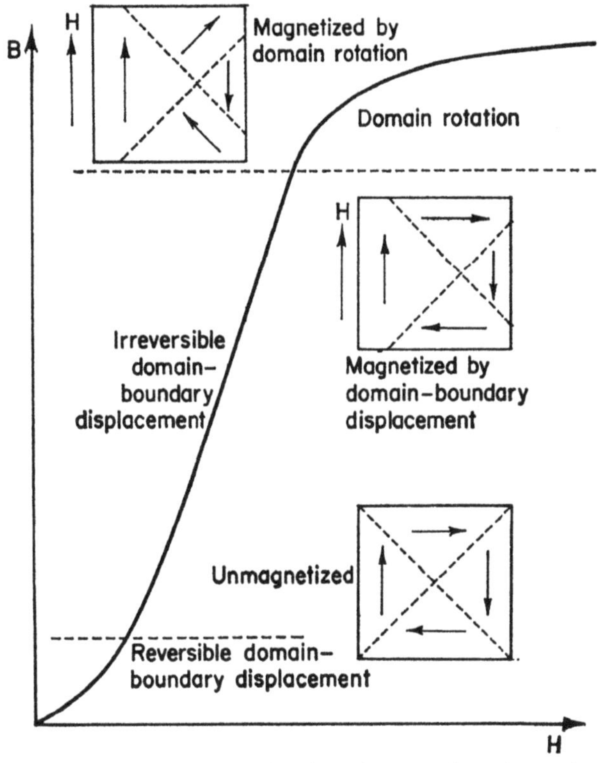

FIG. 6.6. Magnetization by domain growth and rotation.

the domain commences. When all domains have the spin orientation parallel to the external field, saturation is reached. Figure 6.6 illustrates this mechanism. On decrease of the external field, some domains revert to orientations not parallel to the field, but the external field must be reversed to cause the remainder to do so, which leads to the production of the hysteresis loop.

Investigation of the orientation within domains indicates that the crystal lattice behaves in an anisotropic manner with regard to magnetization. The body-centred cubic lattice of iron shows the most ready magnetization in the [100] direction, i.e. along a cube edge, and the most difficult in the [111] direction, i.e. along a cube diagonal. In nickel (f.c.c.) the easy direction is [111] and the hard [100]; in cobalt (c.p.h. lattice) the easy direction is [0001], normal to the close-packed basal plane. Because of this anisotropy, changes in physical dimensions occur on magnetization. Iron crystals expand along the direction of magnetization and contract at right-angles to it, whereas nickel contracts along the magnetization direction and expands perpendicular to it. This phenomenon is known as *magnetostriction*, and is dependent in magnitude on the elastic properties of the crystal in addition to the magnetic ones. Imperfections in the crystal affect the domain growth and particularly the domain rotation in the closing stages of magnetization. If an induction coil connected to an amplifier surrounds the crystal during magnetization, audible clicks occur as domains move—the *Barkhausen effect*.

Magnetic materials can be designed to give properties appropriate to the application. For electromagnetic apparatus (transformers, chokes) a high permeability, giving high saturation values of B, and a low value of coercive force H_C, to reduce hysteresis losses, are required. These are the magnetically soft materials; they are manufactured to high degrees of purity, are generally used in the annealed state, and are preferentially oriented for easy magnetization. Silicon-iron, for instance, used for the laminations of transformers, can be produced with a texture in which the direction of preferred magnetization lies in the rolling direction, or in the rolling plane and normal to the rolling direction. Iron-cobalt alloys have been produced by recrystallizing the sheet in a magnetic field, so that the [110] direction (preferred) sets itself in the direction of the field. The purity is important in these materials, because impurities impede the movement of domain boundaries and the rotation, so increasing H_C and decreasing permeability. Permalloy (Ni_3Fe) is a typical soft magnetic material.

Conversely magnetically hard materials, as required for permanent magnets, must have high values of remanence B_R and of coercive force H_C. The domain-wall movements need to be restrained, and this is achieved by the presence of impurities (particularly those that can be precipitated in the solid), stress, and a fine texture. Typical of the precipitation-hardening materials is Alnico (62% Fe; 26% Ni; 12% Al), which gives a large hysteresis loop.

6.7. *Antiferromagnetism and ferrimagnetism*

Certain substances, usually salts of transition metals, show unusual magnetic behaviour. The exchange energy is negative, and below a critical temperature (the Néel temperature θ_N) the electron spins align antiparallel. If there is a net balance, the material is said to be *antiferromagnetic*.

Metal oxides are often antiferromagnetic; the alignment of metal atom electron spin is counterbalanced by opposite spin in the oxygen atoms. When equal numbers of each spin direction are present, to pair off the electron spins exactly, the material behaves as an antiferromagnetic below the Néel temperature, and as a paramagnetic above it, where the spin arrangement is disordered. A maximum occurs in the susceptibility curve with temperature, at θ_N. Above θ_N, $\chi = C/(\theta + \theta_e)$. Below θ_N, χ is proportional to θ.

The thermal expansion and specific heat of these antiferromagnetic materials also vary with temperature in an anomalous fashion. When the balance between spin directions is not quite equal, but nevertheless there is a fair proportion of antiparallel spin alignment, the material is said to be *ferrimagnetic*. This behaviour is characteristic of the *ferrites* and *spinels* referred to in Chapter 3.

The oxide of iron Fe_3O_4, magnetite, is typical of this class. This oxide has Fe^{2+} and Fe^{3+} ions in the tetrahedral and octahedral voids of an f.c.c. lattice of oxygen ions (see page 58). There are eight Fe^{3+} ions in tetrahedral holes and eight Fe^{3+} plus eight Fe^{2+} ions in octahedral holes in a face-centred cubic cell of 48 closest-packed oxygen atoms. There are thus eight Fe^{3+} ions in octahedral holes, with eight Fe^{3+} ions in tetrahedral holes, whose electrons balance each other in antiparallel spin, as in an antiferromagnetic. This leaves the eight Fe^{2+} ions in octahedral holes to supply the resultant magnetic moment. If the structure were not so balanced by alignment of antiparallel spins, the moment would be much higher. The resistivity of ferrites is about 10^{10} times that of metallic conductors at room temperature, and so induced (eddy) currents cannot circulate in them, which is a great advantage at high frequencies. Mention has already been made of their use in magnetic switching and storage in computers.

The spinel structure can be varied over wide limits, as explained in Chapter 3, and this allows flexibility in the design of materials such as ferrites to vary the magnetic properties. Ferrites give a characteristic *B–H* curve of rectangular shape, with almost instantaneous saturation in the direction of the applied field.

Many metal oxides and sulphides vary in their magnetic behaviour according to composition, usually changing from ferromagnetic to antiferromagnetic behaviour as the pairing of electron spins is changed by the presence of substitutional atoms.

6.8. *Dielectrics*

The behaviour of insulators was referred to in Chapter 2 in dealing with the electronic structure of the solid state. An insulator differs from a semiconductor only in the size of the gap between the valence and conduction band of energies. Even impurities present in insulators have their energy levels so far removed from the required levels that they make little contribution to conductivity either as donor or acceptor atoms. Although insulators do not show much displacement of the positive and negative ions when an electric field is applied, there is obviously some small movement, and the slight separation of the unlike charges leads to local dipoles, and therefore to polarization of the crystal. If the insulator already contains dipoles, as in complex ions or molecules, then the application of an external field will cause these to orientate themselves in the direction of the field. Thermal agitation tends to destroy the orderliness of this polarization, so that the effect is temperature-dependent. Non-polar crystals, having no dipoles in the absence of an external field, do not show this temperature dependence.

There is some similarity between the phenomenon of polarization and that of magnetization. When an electric field affects polarization in a medium, there is a resultant induced effect that modifies the field in a manner similar to the permeability in magnetization. In polarization phenomena, this is known as *permittivity*. The use of insulating materials in capacitors involves such effects, and the *dielectric constant* is defined as the ratio of permittivity of the material to that of a vacuum. The total dipole moment induced in unit volume of the material is called the *polarization density*, and this is determined by three possible processes:

(1) Electronic polarization, produced by relative displacement of electrons and the associated nuclei.
(2) Ionic polarization, produced by similar displacement of positive and negative ions.
(3) Displacement of permanent dipole moments in complex ions or molecules where these exist.

When a polarized material is stressed, there will be a displacement of the two portions of the individual dipoles, and the original balance of moments in the crystal may be altered, particularly if these are not symmetrical. When this occurs, an electric field is set up, and this phenomenon is known as the *piezoelectric effect*. The converse is also true: that an electric field applied to such a crystal causes dipole rearrangement, and a mechanical strain results. Such piezoelectric crystals can act as mechanical–electrical transducers, and are utilized in microphones, vibrators, and gramophone pick-ups for this reason. Similar changes can be produced by heating polarized structures, the effect known as *pyroelectricity*.

For a crystal to show the piezoelectric effect, the dipoles must be permanently aligned in the structure, and not require the application of an external field. The same argument applies to pyroelectricity. When an external field is applied, some crystals permanently reverse the direction of polarization, and these are referred to as *ferro-electric*. Ferroelectricity is the electrical analogue of ferromagnetism. There is also a characteristic temperature, above which the dielectric constant drops suddenly and the permanent polarization is changed; this temperature is called the *Curie point*. Ferroelectric domains exist, as in ferromagnetism, consisting of areas of crystal lattice in which the direction of polarization changes. A distortion occurs in ferroelectrics similar to magnetostriction, and is called *electro-striction*. Whereas magnetostriction depends on the relative field direction, electrostriction is determined by the crystal structure. Hysteresis loops are found when ferroelectrics are subjected to an alternating field, as would be expected.

Application of ferroelectrics includes memory units in computers and in capacitors. A number of phosphate salts (Rochelle salt) are ferroelectric, and another typical example, on which much structural work has been done, is barium titanate $BaTiO_3$ (see Chapter 3). Salts of amphoteric oxides of transition metals often fall into this class, e.g. niobates ($NiNbO_3$) and tantalates, and tungstic oxide WO_3.

6.9. *Optical properties*

The index of refraction in a material μ is defined as the ratio of the velocity of light in vacuum c to the velocity in the medium v

$$\mu = c/v$$

The velocity of light in a medium v is also determined by the dielectric constant

$v = c/\sqrt{K_e}$ where K_e = dielectric constant so that $\mu = \sqrt{K_e}$

This relationship is not valid when the dielectric constant varies with frequency, but is true over the frequency range in which the material is transparent to radiation. For most ionic and molecular materials (solutions and gases) it holds in the visible spectrum.

At higher frequencies, absorption of electromagnetic radiation commences, and at still higher values, reflection may occur.

In metals, the electron cloud permeating the lattice absorbs incident light with electrons excited to higher energy levels in the conduction band. The modern zone theory that deals with these transitions is outside the scope of this book. A minimum absorption frequency can be calculated and from this an estimate can be made of the wavelength of radiation which will be absorbed. For sodium, this is 6,200 Å, for rubidium 11,000 Å, for copper 3,700 Å, and for silver 4,700 Å. These figures indicate that the alkali metals absorb light even in the longer-wavelength infrared, whereas copper and silver do so in the visible or near ultraviolet region, of shorter wavelength. Such deductions are confirmed in principle by experiment, especially for the alkali metals. Very thin foils have to be used in such experiments, but sodium, for example, is transparent to visible light when of the order of 10^{-3} mm thick. The colour of metals is a consequence of their absorption properties. Incident light of mixed wavelength is mostly reflected from polished metallic surfaces, but the absorption that takes place in the surface layers is selective in wavelength as shown above, and the intensity of these wavelengths is reduced in the reflected ray. Consequently, the colour of a metal is roughly the light remaining after absorption, the complementary colour, in fact. Copper begins to absorb at wavelengths of the order of 5,000 Å; much of the blue light is absorbed, and reddish light is reflected. Gold, similarly, cuts out the blue light, but more of the yellow is reflected than in the case of copper. Silver also cuts out some blue, but much less than copper or gold, and so the reflected light is white, but with a higher proportion of yellow than, say, chromium. Most other metals appear white because absorption commences at the blue end of the spectrum or in the ultraviolet. When gold is prepared in thin foils, the transmitted light can be seen; the colour is, of course, complementary to that of the reflection, i.e. greenish-blue.

Copper, silver, and gold are all good conductors, and all contain ten d electrons. Nickel, preceding copper in the periodic table ($(3d)^8$ electrons), also has yellowish tints in reflected light. Electrical conductivity must obviously play a part, since the transition of electrons to higher levels is involved, exactly as in conductivity. When alloyed together, copper, silver, and gold retain the yellow colour in the main, so that the absorption must be altered only slightly. Electrical conductivity is only slightly reduced in the alloys of these three metals, the electron transitions are not violently affected, and absorption of light suffers only minor effects. When they are alloyed with other metals, however, particularly those that influence the crystal lattice and therefore the electron-cloud distribution, the colour changes are more obvious. Zinc affects the colour of copper when present above 5%, but the colour finally disappears only at 50% zinc. Tin affects the colour of copper also above 5%, but destroys it completely at 30%. Some alloys of copper are unusual, one with antimony is a beautiful purple.

Semiconductors behave exactly as one would expect from the electron-transition model. At longer wavelengths (far infrared), they are more or less transparent. The absorption coefficient frequently decreases as the wavelength shortens, showing that electron transition to higher levels is comparatively rare. At a particular frequency, however, there is a sudden rapid increase in absorption (in germanium, for instance, at about 2×10^4 Å) which corresponds in energy to the gap between the valence and conduction bands. Thus, the semiconductor is suddenly able to undergo electron transitions with ease, the photons of radiation having exactly the appropriate energy. The energy is absorbed by the transition, and the material becomes opaque. As the wavelength is lowered further, the electron transitions can proceed into excited states, within the conduction band. The subsequent return of some of the electrons into lower energy states gives rise to emission, as with metals, and the material becomes reflective, with a metallic appearance. Any increase in temperature, which helps to supply part of the energy for electrons to bridge the gap, is bound to have an effect on the absorption of light.

Insulators, having large energy gaps in the forbidden regions, are transparent to visible light, because the photons cannot bridge the gap. According to the theory discussed above, they should, however, become opaque at sufficiently high energies, i.e. with radiation of very short wavelength. In fact, they do show absorption in the ultraviolet, but not in the same manner as semiconductors. Instead, a series of peaks occurs in the measured absorption coefficient, varying with the material and with wavelength. The explanation is that a different mechanism is in operation. With a semiconductor, the absorption commences once the gap between valence and conduction band is bridged, in a similar way to the increase in conductivity with rise in temperature. Such semiconductors become photoconductive at this stage, i.e. the incident light produces an effect similar to the rise in conductivity with temperature. This does not happen with insulators, so that simple polarization is not responsible, nor a simple transition of electrons bridging the gap.

The explanation of the peaks was first put forward by FRENKEL, referring to alkali halide crystals. It is assumed that an electron can be excited out of the valence band to an energy state not much above that band, and actually in the forbidden region between valence and conduction bands. This is possible by forming an electron-hole pair, called an *exciton*. The Coulomb attraction of these two opposite charges holds the pair together. It is modified by the dielectric constant of the material, and is inversely proportional to the separation distance between electron and hole. The binding energy must, of course, be less than the gap in energy between the two bands. These excitons absorb energy, and so contribute to the absorption of incident electromagnetic radiation. Temperature affects the exciton formation, by virtue of the thermal vibration of the atoms, and the excitons can move through the crystal over short distances. They can, of course, recombine at any stage, but since they are electrically neutral, no current flows as a result of their motion or recombination.

In addition to the alkali halides, cuprous oxide Cu_2O also shows similar behaviour, but other ionic crystals differ in the detail of their absorption. Molecular crystals, having homopolar bonds, show sharp absorption peaks and are believed to undergo similar exciton formations.

Window glass transmits visible light well, and the ultraviolet to some extent, but is opaque to infrared. It can therefore be used as a fire-screen, cutting out the infrared. Similarly, in the hot-house the glass transmits the visible light and some of the ultraviolet, and this warms up the soil. The radiation re-emitted from the soil and from the inside structure is of longer wavelength (infrared), as one would expect from a low-temperature source of radiation. The glass does not transmit this but absorbs it, so retaining the heat inside. In contrast, quartz is transparent to the long-wavelength infrared, opaque to short infrared, and transparent again in the ultraviolet. It is therefore used in spectrometers as the dispersion prism. Rock-salt and fluorite both transmit the infrared very well; fluorite also transmits ultraviolet.

Water vapour in the earth's atmosphere cuts out much of the infra-red from the sun, and a high concentration of carbon dioxide can do likewise. Certain applications of radiation to the measurement of temperature may suffer as a result, e.g. in measuring the temperature (by total radiation) of a hot body inside a furnace fired by gas and containing carbon dioxide as a combustion product.

The transparency of insulators to visible light is sometimes accompanied by the appearance of colours in the crystals. The colour can arise only from selective absorption of certain wavelengths, as with metals, and this is believed to arise from impurities in the crystal. These imperfections act in a similar way to impurities in semiconductors, by providing a series of local energy levels in the forbidden region between the valence and conduction band. Photons of energy far too small to bridge the forbidden gap may be affected by absorption in these intermediate levels in the impurities. Many of the alkali halide crystals have been studied in this way, and the impurities referred to as colour centres or *F centres* (German *Farbe* colour). The presence of excess metal ions in the non-stoichiometric crystal also leads to such colour centres. Niobium in its fluoride crystal gives a pink coloration; excess potassium in KCl makes it blue; and excess sodium in common salt gives a yellow colour. Many gemstones are coloured through the same agency, particularly when different forms of the same mineral are involved.

The energy levels associated with these F centres are frequently close to the lower edge of the conduction band, and thermal excitation by rise in temperature may allow bridging of this smaller gap. If this occurs, the crystal should become photoconductive; this has been observed. Many other types of transition have been demonstrated, similar to the formation of F centres, and it is quite likely that applications may be developed for these new materials in the near future.

Photographic processes involve similar phenomena to F-centre formation, except that metallic silver actually precipitates out of the bromide crystal, and growth of these precipitates is rapid. Colour film is dependent on the selective absorption produced by the addition of dyes, which affect the precipitation of the latent image.

The absorption of light by electron transitions between different energy levels may be followed by return of the electron to lower levels, and re-emission of light. This property is called *luminescence*. The time-delay varies considerably, and serves to classify substances into two main groups. *Fluorescent* materials are those in which the emission takes place less than 10^{-8} seconds after absorption. *Phosphorescent* materials emit light after an interval greater than 10^{-8} seconds, and their emission lasts much longer. There is considerable evidence to show that luminescence depends on the presence of impurities, called *activators*. The emitted light is always of longer wavelength, and so is not reabsorbed in the material. The emission comprises also a band of wavelengths and not single ones, because of the thermal motion of the atoms during emission. Zinc sulphide is a typical phosphor, as are the organic materials camphor and anthracene. Their uses are obvious in illumination of instrument and clock dials, and for safety indicators in dark conditions.

The most recent phenomenon of electromagnetic radiation absorption is the discovery of *light amplification* by *stimulated emission* of *radiation*, or *laser*, operation. Laser operation was discovered in ruby in 1960. Ruby is primarily crystalline aluminium oxide Al_2O_3 in which chromium ions replace some of the aluminium ones, giving the ruby colour. The electrons of the chromium are normally in the ground state, but can be excited by irradiation with green light, which they absorb readily. They increase in energy, but may fall back fairly readily into an intermediate state, and from there to the ground state again. The emission will be fluorescent radiation, in the red end of the spectrum. When the crystal is irradiated with the red light that the chromium atoms are about to emit, it is found that a much more rapid emission takes place than would normally occur. The

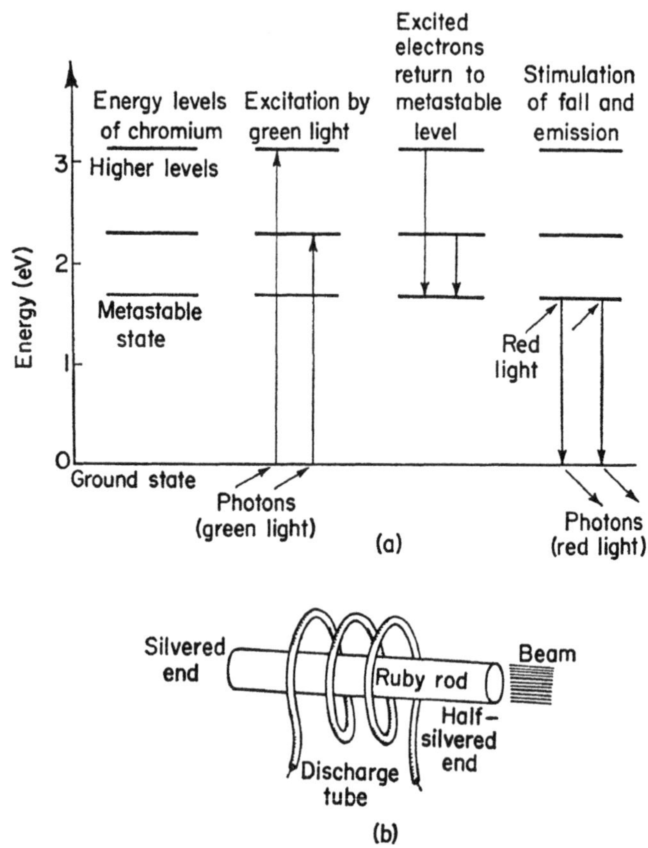

FIG. 6.7. Laser operation. (*a*) Excitation into higher states and return by stimulation. (*b*) Practical apparatus.

stimulated emission is much more intense and more rapid than the normal fluorescence that results. The laser apparatus consists simply of a rod of ruby, half silvered at the emission end face and fully silvered at the other, and surrounded by a discharge tube emitting a broad band of wavelengths. The green light from the discharge tube excites the electrons in chromium, and the red light stimulates the emission as the electrons fall in energy level. The stimulated emission from one end of the rod will continue to stimulate other atoms, until a cascade of red light leaves the half-silvered end of the rod. Because of the silvering at the ends, and reflection from the sides back into the crystal, a powerful parallel beam emerges. The pulse lasts for about 5×10^{-4} sec, at a power output of about 10 kW, with a beam cross-sectional area of 1 cm^2. Continuous operation of lasers is being developed, and high-temperature melting operations have been carried out by this means. A similar type of stimulated emission can occur at microwave frequencies, where the magnetic energy levels of paramagnetic ions in crystals are concerned, and these devices are called *masers*. Figure 6.7 illustrates the laser action.

6.10. *Other electromagnetic radiations*

Short radio waves (10^6–10^{11} Å) are readily absorbed by metals; e.g. tinfoil only 0·01 mm thick completely prevents transmission. Dielectrics, however, are completely transparent to these radiations. Substances such as sulphur and ebonite (rubber hardened by mixing with sulphur and heating) are transparent to wavelengths from 5 to 20 cm. Glass and wood, normally regarded as insulators, are not so transparent to radio waves. For reasons such as those described, radio reception inside a metal box (e.g. an automobile) is useless without an external aerial, but is reasonable inside a brick house, becoming less so if a wooden box surrounds the aerial.

Metals are not opaque to lower wavelengths than the ultraviolet, e.g. X-rays, from 10^{-1} to 10^3 Å. The frequency of these radiations is so high that the electron cloud has little effect. With γ-rays, down to 10^{-2} Å, the penetration through metals is even greater; hence the use of X- and γ-rays for radiography. X-rays are, of course, affected by the ions and the innermost energy levels of electrons. Soft X-rays can be used for investigation of the electronic configurations, and hard X-rays for reflections from atoms to investigate the lattice type and interatomic distance.

7

Mechanical Properties

7.1. *General aspects*

From the point of view of mechanical properties the use of engineering materials may involve three distinct possibilities:

(1) High resistance to elastic deformation, and the deliberate avoidance of plastic deformation and fracture. The majority of highly stressed machine components and many large structures such as bridges and buildings come into this category.
(2) Low resistance to elastic and plastic deformation but deliberate avoidance of fracture. Processes of metal-working by rolling, forging, extrusion, drawing, and stretching are involved in this application.
(3) Rapid elastic and plastic deformation leading to fracture. All machining processes, involving the actual removal of material, are involved, whether by sharp-edged tools or abrasion.

All three processes are concerned with fracture mechanisms, the first two in avoiding these, the last seeking to achieve fracture efficiently and economically. The first two require a detailed knowledge of plasticity, the first to avoid plastic deformation, the second to achieve it economically without the final stage to fracture. Only the first is primarily concerned with elasticity, and even then the concern is rather more with the transition from elastic to plastic behaviour than the elastic range itself.

7.2. *Elasticity*

There are two approaches possible. The first is that applied to engineering design, and the second that applied to the study of the properties of materials. The elastic theory applied to single crystals is not normally of significance to engineering design, which is concerned more with bulk properties in polycrystalline materials. There is an increasing tendency, however, to relate the basic elastic theory to the behaviour of the polycrystal, because similar arguments can sometimes be used in the extension to plastic behaviour.

When a tensile or compression load is applied to a material, the *normal stress* σ is defined as the load divided by the area normal to the load over which the load is acting. The *strain* ε which results from this stress is defined as the change in length per unit length of the specimen. The relationship between the two is a constant for small strains, where the conditions are said to be truly elastic

$$\sigma = E\varepsilon$$

where E is a constant, Young's modulus of elasticity. The proportionality of stress and strain was first recognized by HOOKE. YOUNG appreciated that the constant of proportionality is a regular property of the material under test.

If a stress is applied tangentially through a plane in the material (Figure 7.1), a lateral strain results by displacement of this plane relative to other planes parallel to it. Such *tangential stresses* are

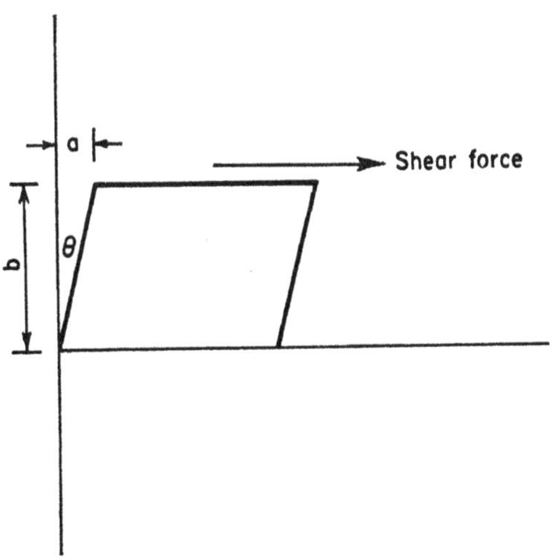

FIG. 7.1. Shear strain.

frequently referred to as *shear*, and the strains as *shear strains*. The shear stress τ is related to the shear strain γ by a similar proportionality to the normal stress and strain,

$$\tau = G\gamma$$

where G is a constant, the modulus of rigidity, or shear modulus. The shear strain can be measured as the tangent of the angle through which the planes are displaced relative to each other. If the planes are distant b apart, and the relative displacement is a, then

$$\tan \theta = a/b = \gamma$$

Figure 7.2 represents a cube of material that is being uniformly stressed. There are three couples acting in the x direction. The first

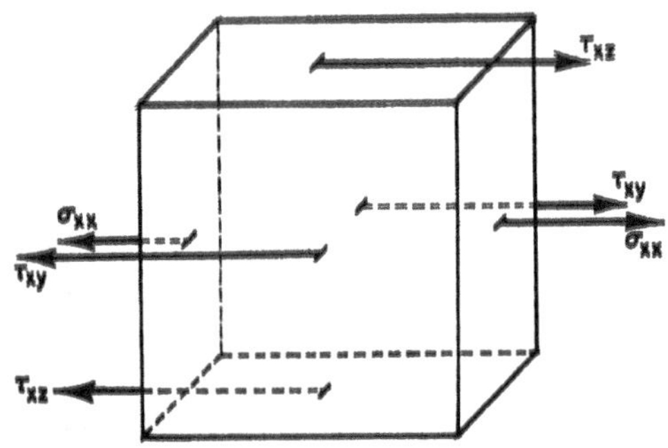

FIG. 7.2. Elastic stress components in the x direction.

is applied to faces perpendicular to the x axis, and the normal stress denoted by σ_{xx}, positive in tension. The second couple is applied to faces perpendicular to the y axis, and the shear stress is given the symbol τ_{xy}. The third couple applies to faces perpendicular to the z axis, and the shear stress is τ_{xz}. Similarly, three couples act in the y direction, a normal stress σ_{yy} (face perpendicular to the y axis) and shear stresses τ_{yz} and τ_{yx} (faces perpendicular to z and x axis respectively). Finally, three couples in the z direction are designated σ_{zz}, τ_{zx}, τ_{zy}. These components can be combined in matrix form:

$$\begin{bmatrix} \sigma_{xx} & \tau_{xy} & \tau_{xz} \\ \tau_{yx} & \sigma_{yy} & \tau_{yz} \\ \tau_{zx} & \tau_{zy} & \sigma_{zz} \end{bmatrix}$$

The components can be reduced, by virtue of the similarity between some of the shear stresses. τ_{yx} produces the opposite effect to τ_{xy}, and so the two cancel out. The same applies to $\tau_{xz} = \tau_{zx}$ and $\tau_{yz} = \tau_{zy}$. This leaves us with six components, σ_{xx}, σ_{yy}, σ_{zz}, τ_{xy}, τ_{yz}, τ_{zx}.

These stresses produce deformation, and six corresponding strain components make up the strain tensor, denoted by three normal strains ε_{xx}, ε_{yy}, and ε_{zz}, and three shear strains γ_{xy}, γ_{yz}, γ_{zx}. The linear relations between stress and strain are given by the six equations:

$$\begin{aligned}
\sigma_{xx} &= c_{11}\varepsilon_{xx} + c_{12}\varepsilon_{yy} + c_{13}\varepsilon_{zz} + c_{14}\gamma_{yz} + c_{15}\gamma_{zx} + c_{16}\gamma_{xy} \\
\sigma_{yy} &= c_{21}\varepsilon_{xx} + c_{22}\varepsilon_{yy} + c_{23}\varepsilon_{zz} + c_{24}\gamma_{yz} + c_{25}\gamma_{zx} + c_{26}\gamma_{xy} \\
\sigma_{zz} &= c_{31}\varepsilon_{xx} + c_{32}\varepsilon_{yy} + c_{33}\varepsilon_{zz} + c_{34}\gamma_{yz} + c_{35}\gamma_{zx} + c_{36}\gamma_{xy} \\
\tau_{yz} &= c_{41}\varepsilon_{xx} + c_{42}\varepsilon_{yy} + c_{43}\varepsilon_{zz} + c_{44}\gamma_{yz} + c_{45}\gamma_{zx} + c_{46}\gamma_{xy} \\
\tau_{zx} &= c_{51}\varepsilon_{xx} + c_{52}\varepsilon_{yy} + c_{53}\varepsilon_{zz} + c_{54}\gamma_{yz} + c_{55}\gamma_{zx} + c_{56}\gamma_{xy} \\
\tau_{xy} &= c_{61}\varepsilon_{xx} + c_{62}\varepsilon_{yy} + c_{63}\varepsilon_{zz} + c_{64}\gamma_{yz} + c_{65}\gamma_{zx} + c_{66}\gamma_{xy}
\end{aligned}$$

The coefficients c_{ij} are the moduli of elasticity referring to particular crystallographic directions, but the matrix of c_{ij} is symmetrical, and the number can be immediately reduced to 21 independent components. In addition, the symmetry of the crystal affects the coefficients, and the number may be reduced even further,

Isotropic crystals:	number of independent coefficients = 2
Cubic crystals:	number of independent coefficients = 3
Hexagonal crystals:	number of independent coefficients = 5
Tetragonal crystals:	number of independent coefficients = 6
Orthorhombic crystals:	number of independent coefficients = 9

For cubic crystals, we are left with:

$$\begin{bmatrix}
c_{11} & c_{12} & c_{12} & 0 & 0 & 0 \\
c_{12} & c_{11} & c_{12} & 0 & 0 & 0 \\
c_{12} & c_{12} & c_{11} & 0 & 0 & 0 \\
0 & 0 & 0 & c_{44} & 0 & 0 \\
0 & 0 & 0 & 0 & c_{44} & 0 \\
0 & 0 & 0 & 0 & 0 & c_{44}
\end{bmatrix}$$

For hexagonal crystals

$$\begin{bmatrix}
c_{11} & c_{12} & c_{13} & 0 & 0 & 0 \\
c_{12} & c_{11} & c_{13} & 0 & 0 & 0 \\
c_{13} & c_{13} & c_{33} & 0 & 0 & 0 \\
0 & 0 & 0 & c_{44} & 0 & 0 \\
0 & 0 & 0 & 0 & c_{44} & 0 \\
0 & 0 & 0 & 0 & 0 & \frac{1}{2}(c_{11} - c_{12})
\end{bmatrix}$$

For quartz, a trigonal symmetry,

$$\begin{bmatrix} c_{11} & c_{12} & c_{13} & c_{14} & 0 & 0 \\ 0 & c_{11} & c_{13} & -c_{14} & 0 & 0 \\ 0 & 0 & c_{33} & 0 & 0 & 0 \\ 0 & 0 & 0 & c_{44} & 0 & 0 \\ 0 & 0 & 0 & 0 & c_{44} & 0 \\ 0 & 0 & 0 & 0 & 0 & \frac{1}{2}(c_{11} - c_{12}) \end{bmatrix}$$

Table VI gives some values of these elastic moduli for cubic crystals at room temperature.

TABLE VI

Elastic constants for some cubic crystals at room temperature
(Units of 10^{11} N m^{-2} or 10^{8} dyne m^{-2})

Material		Crystal lattice	c_{11}	c_{12}	c_{44}
Iron	Fe	b.c.c.	2·37	1·41	1·16
Tungsten	W	b.c.c.	5·01	1·98	1·15
Copper	Cu	f.c.c.	1·684	1·214	0·75
Diamond	C		9·2	3·9	4·3
Silicon	Si	f.c.c.	1·66	0·64	0·79
Germanium	Ge	f.c.c.	1·29	0·48	0·67
Aluminium	Al	f.c.c.	1·08	0·62	0·28
Lead	Pb	f.c.c.	0·48	0·41	0·14
Sodium chloride	NaCl		0·486	0·127	0·128

In general, the highest value of Young's modulus occurs in the direction of closest-packed atoms.

In practice, the mechanical properties of engineering materials are mainly of interest for crystals of high symmetry, and the constants used are related more to the methods of testing materials.

For orthorhombic materials, the normal stress σ_{xx}, applied in the x direction, is regarded as producing a strain ε_{xx} in the x direction, and strains $-\varepsilon_{yy}$ and $-\varepsilon_{zz}$ in the transverse directions y and z. The relationships may be written:

$$\sigma_{xx} = E_1 \varepsilon_{xx}$$

$$\varepsilon_{yy} = -\mu_{12}\varepsilon_{xx} = -\mu_{12}\sigma_{xx}/E_1$$

$$\varepsilon_{zz} = -\mu_{13}\varepsilon_{xx} = -\mu_{13}\sigma_{xx}/E_1$$

where μ_{12} and μ_{13} are called *Poisson's ratios*, the ratio of deformation in one direction to that in a perpendicular direction. Similar expressions can be found for σ_{yy} and σ_{zz} in terms of similar constants.

$$\sigma_{yy} = E_2\varepsilon_{yy}$$
$$\varepsilon_{zz} = -\mu_{23}\varepsilon_{yy} = -\mu_{23}\sigma_{yy}/E_2$$
$$\varepsilon_{xx} = -\mu_{21}\varepsilon_{xx} = -\mu_{21}\sigma_{yy}/E_2 \quad \text{etc.}$$

The reciprocity of relationships noted for single crystals applies again, such that $E_1/\mu_{12} = E_2/\mu_{21}$; $E_2/\mu_{23} = E_3/\mu_{32}$; $E_3/\mu_{31} = E_1/\mu_{13}$. Hence, there are three values for Young's modulus, E_1, E_2, E_3, and three independent values of Poisson's ratio, μ_{12}, μ_{23}, μ_{31}. A shear stress τ_{yz} in the yz plane produces a shear strain γ_{yz}, and hence

$$\tau_{yz} = G_1\gamma_{yz}$$

where G_1 is a modulus of rigidity. Similar reasoning indicates three moduli, G_1, G_2, G_3. These are the same constants as c_{44}, c_{55}, c_{66}, discussed above.

An orthorhombic crystal can thus be defined by nine independent constants, three elastic moduli, three moduli of rigidity, and three values of Poisson's ratio. A tetragonal crystal involves only six independent constants, since two sides of the unit cell are equal, so that $\mu_{12} = \mu_{13}$; $E_2 = E_3$; $G_2 = G_3$. For cubic crystals, all three sides of the unit cell are equal, and so only three constants remain, E; G; μ.

In a perfect isotropic body, the coordinate axes can be chosen so that the shear stresses on all faces of the cube are zero, and then stresses normal to cube faces are termed *principal stresses*, designated σ_1, σ_2, σ_3. Only two independent constants are now required, E and μ, and

$$\begin{aligned} \varepsilon_1 &= 1/E[\sigma_1 - \mu(\sigma_2 + \sigma_3)] \\ \varepsilon_2 &= 1/E[\sigma_2 - \mu(\sigma_1 + \sigma_3)] \\ \varepsilon_3 &= 1/E[\sigma_3 - \mu(\sigma_2 + \sigma_1)] \end{aligned} \tag{7.1}$$

A polycrystalline cubic metal behaves as an isotropic body, and the relation between G, E, and μ is

$$G = E/2(\mu + 1) \tag{7.2}$$

Isotropic pressure is equivalent to three equal and normal tensile stresses acting in three mutually perpendicular directions. The bulk modulus K is defined as the ratio of isotropic pressure to the relative volume change produced by that pressure. For cubic materials, each of the three principal stresses produces a relative elongation

$$\varepsilon = -P/E + \mu P/E + \mu P/E$$

where P is the principal stress; whence

$$\varepsilon = (2\mu - 1)P/E$$

The relative volume change is three times the elongation, i.e. 3ε.

Hence $$\Delta v/v = 3\varepsilon = (2\mu - 1)3P/E$$

$$K = -\frac{P}{\Delta v/v} = \frac{-PE}{(2\mu - 1)3P}$$

$$= \frac{E}{3(1 - 2\mu)} \tag{7.3}$$

and since $$G = \frac{E}{2(\mu + 1)}$$

$$K = \frac{2G}{3}\frac{(1 + \mu)}{(1 - 2\mu)} \tag{7.4}$$

Attempts have been made in the past to simplify even further the elastic constants of the general tensor, but too many assumptions are involved for the results to be universally applicable. It is, however, possible to estimate the probable values of μ for different lattices. Homopolar or covalent bonds should give rise to a value of $G > c_{12}$, and $\mu < 0{\cdot}25$. The ionic bond and Van der Waals' bonding should give $c_{12} = c_{66} = G$, and $\mu = 0{\cdot}25$. Metallic bonding should give $\mu > 0{\cdot}25$, $(c_{12} > G)$, and long-chain molecules $\mu \gg 0{\cdot}25$. There is a maximum value of μ of 0·5, since K (equation (7.3)) can never be negative. The measured values of μ for isotropic substances confirm this argument.

Material	μ
Silica SiO_2	0·16
Glass	0·2
Steel	0·27
Nickel	0·31
Aluminium	0·32
Copper	0·34
Lead	0·43
Rubber	0·50

For the three main classes of materials, the relationships between the elastic moduli are roughly as follows:

Covalent and amorphous structures	$\mu \simeq \frac{1}{4}$; $G \simeq \frac{2}{5}E$; $K \simeq \frac{2}{3}E$
Metallic lattices and alloys	$\mu \simeq \frac{1}{3}$; $G \simeq \frac{3}{8}E$; $K \simeq E$
Long-chain molecules	$\mu \simeq \frac{1}{2}$; $G \simeq \frac{1}{3}E$; $K \gg E$

Theoretical calculations to predict the moduli from the forces of attraction and repulsion have been moderately successful. Grüneisen first showed that

$$KV_0 = mnU_0/9 \tag{7.5}$$

where K is the bulk modulus, m, n are the constants of the Mie equation for potential energy (see page 29), V_0 is the equilibrium value for molecular volume, and U_0 the equilibrium value for potential energy for one gramme-molecule. Molecules involving binding by Van der Waals' forces have values of $m = 6, n = 9$ (see page 40), and U_0 approximates to the energy of sublimation. This gives K the right order of magnitude for these molecules. In ionic bonding the value of U_0 must be greater than the sublimation energy, because sublimation involves the whole molecule, whereas the lattice is made up of ions. By adding the individual energy levels together, a better value of U_0 can be obtained.

For sodium chloride $U_0 \simeq 760$ kJ/mole

$$n = 9; \quad m = 1; \quad V_0 = Nd^3$$

and $K \simeq 10^{11}$ newtons per square metre (N m^{-2})

For metals, the lattice energy is chiefly derived by the ionization of the metal atoms involved in the metallic bond. Addition of I, the ionization energy, to the sublimation energy, gives a better figure.

For potassium,

Sum of the energy terms $\simeq 550$ kJ/mole

Putting $m = 1; n = 2$ (alkali metal)

$$K \simeq 1{\cdot}3 \times 10^9 \text{ N m}^{-2}$$

This is low by a factor of 4 or so, but not too poor an estimate considering the assumptions made.

The velocity of sound is closely related to elastic moduli, and its measurement provides a most important technique in experimental determination of the moduli. A longitudinal wave, propagated through a bar, gives rise to transverse waves by virtue of Poisson's ratio. The ratio of the velocity of sound for transverse and longitudinal waves can be compared as for Poisson's ratio, between three classes of structure:

Covalent bonds	$\mu \simeq 0{\cdot}25$	$c_t/c_l \simeq 0{\cdot}63$
Metallic bond	$\mu \simeq 0{\cdot}33$	$c_t/c_l \simeq 0{\cdot}61$
Long-chain polymers	$\mu \simeq 0{\cdot}5$	$c_t/c_l \simeq 0{\cdot}58$

High values of the velocity of sound occur in ionic crystals of high valency and in metals of high modulus of rigidity.

When a material is subjected to vibration, the alternating driving force is out of phase (ahead) with the displacement which occurs in the material, and this phase difference is called the *angle of loss* δ. The value of δ is a measure of the damping capacity of the material. If the stress applied is σ, and the displacement ε, then

$$\sigma = \sigma_m \sin \omega t$$

and
$$\varepsilon = \frac{\sigma_m}{E} \sin(\omega t - \delta)$$

where σ_m is the maximum value of stress, E is Young's modulus, ω the frequency, and t the time. These equations give rise to an elliptical figure when σ is plotted against ε (Figure 7.3). The area of the ellipse is a measure of the energy dissipation per cycle. If $\bar{\varepsilon}$ is the amplitude, then the relative decrease in amplitude $d\bar{\varepsilon}/\bar{\varepsilon}$ is equal to $\pi\delta$. The value $d\bar{\varepsilon}/\bar{\varepsilon}$ is called the *logarithmic decrement*, where

$$\bar{\varepsilon} = \bar{\varepsilon}_m \exp(-\pi\delta t/T)$$

$\bar{\varepsilon}_m$ being the maximum amplitude and T the time for one cycle. The time required for the amplitude to diminish to $1/l$ of its original

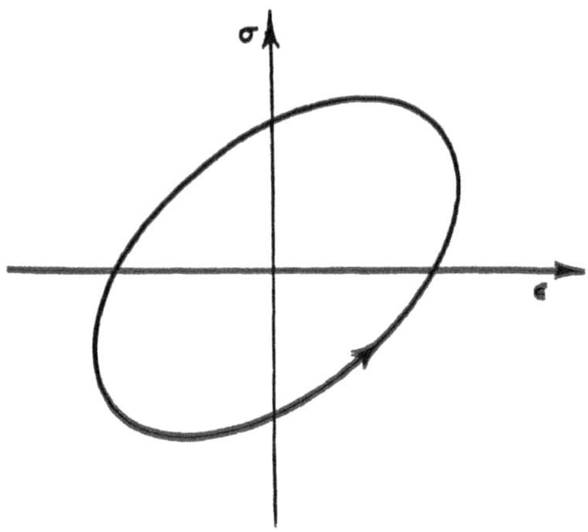

FIG. 7.3. The phase difference between $\varepsilon + \sigma$.

value is called the *relaxation time* t_r, and the distance l over which this occurs is called the *relaxation distance*.

This damping of vibrations, in the absence of any external forces, is dependent solely on the internal friction of the material. For metals, δ is of the order of 10^{-3}–10^{-4}, for quartz about the same. Brick and concrete show angles of loss of about 10^{-1}, wood about 1, cork 3 to 4, and rubber 5 to 6. The presence of impurities, internal stresses and grain boundaries affects the value markedly. Impurities increase δ, as do internal stresses from cold work or heat treatment. The greater the grain size of a material (fewer grain boundaries) the larger the angle of loss. The more complex the crystal lattice, the smaller the angle δ. Rise in temperature increases δ, largely owing to effects at the grain boundaries. An increase of amplitude of vibration also tends to increase δ. The effect of ferromagnetism is marked, damping being higher in the unmagnetized or only partially magnetized state.

Damping is related to the after-effect of elastic stress. When a load is applied to a body, an instantaneous strain results, which increases slowly over a period of time without further increase in load. The rate of this increase diminishes, and movement finally ceases. On removal of the load, an instantaneous contraction first results, and then precisely the same pattern of further contraction occurs at a decreasing rate until the original dimensions are restored (Figure 7.4). This phenomenon is known as the *elastic after-effect*. Between 25 and 90% of the total deformation (in different materials) is contributed from this source. The after-effect occurs partly from the slow evolution of heat due to the stress, and the effect of this heat on thermal expansion; partly from the diffusion of impurity atoms

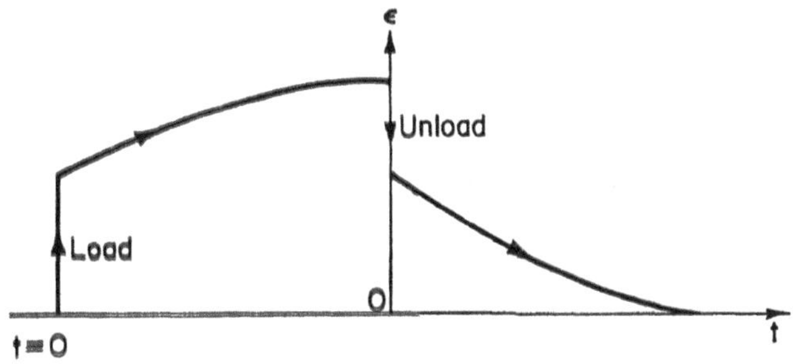

FIG. 7.4. Elastic after-effect on loading and unloading.

in pure materials or absorbed molecules in porous materials; and partly from mechanical hysteresis effects of the grains. Single crystals do not show an after-effect.

The presence of impurity atoms in metals causes many effects, some due to diffusion from initial lattice sites into those giving rise to less strain (Snoek). This type of influence is temperature-dependent, for at low temperatures the diffusion rate is too slow, and at high temperatures it can occur too quickly to be a retarding effect.

Similar features apply to non-metallic materials. Dislocations have a pronounced effect in all materials, since they affect diffusion processes and the movements of atoms. The measurement of internal friction, and the separation of these individual contributory factors, has proved of enormous value in the study of deformation mechanisms and the fine structure of materials, particularly the metals.

Components whose damping is required to be small, e.g. cymbals, bells, are made of materials of low damping capacity, of which alloys of copper are good examples. Components requiring high damping, e.g. anti-vibration mountings, use rubber, cork, or similar porous materials such as balsa wood and foamed polymers.

7.3. *Plasticity*

The theory of plasticity, as used in engineering design studies, is continually undergoing improvement, and is rather outside the scope of this book. The various processes in which plasticity plays a large part can now be appreciated, however, as a result of developments in dislocation theory. Although this work has concentrated mainly on the metals, there is a growing tendency to introduce it in the study of metallic oxides, partly as a result of their increasing application, and partly because of the particular behaviour of their crystal lattices.

In Chapter 3 the properties of those crystal lattices most frequently found in engineering materials were described, and of the two main processes of deformation, slip and twinning, were mentioned. Chapter 4, dealing with the presence of defects, indicated the modification to the slip process introduced by the presence of dislocations. Many materials show only a very short elastic range and begin to deform plastically at comparatively low stresses. As explained in Chapter 4, this may be due to the smaller atomic movements required in plastic flow, when dislocations are present in large numbers. Temperature has a pronounced effect on the point at which elastic deformation changes to plastic, referred to as the *elastic limit*. The effect of temperature may be quite complex, for thermal agitation not only assists the stress, to cause deformation more readily, but also increases diffusion of atoms of all kinds and leads to changes in the operative slip or twinning systems. This becomes very important in high-temperature service, involving the process known as *creep*.

Although the elastic limit is strictly the point at which irreversible (plastic) deformation commences, it is not easy to observe precisely. At a somewhat higher stress, many materials suddenly commence to flow plastically at a higher rate, and this point is referred to as yielding, and the stress level as *yield* or *flow stress*. Some materials behave in an unusual manner, in that the stress drops at this stage, and flow proceeds suddenly without any increase in stress for a time, although later the stress must rise to continue the deformation. These phenomena can be described on the basis of dislocation theory.

7.3.1. *Resolved shear stress*

When a tensile or compressive stress is applied to a specimen of regular shape, such as a cylinder, slip will commence only where a particular glide plane and glide direction are so oriented to the stress system that atomic movement by slip or twinning can take place (Figure 7.5). The stress system can be interpreted best in terms of the shear stresses acting on the slip plane in the particular slip direction, since the movement of one lattice plane of atoms (or a portion of it) over another is in fact a shear phenomenon. The applied shear stress σ necessary to start the slip process is given by

$$\tau = \sigma \cos \phi \cos \lambda \tag{7.6}$$

where τ is the critical shear stress for slip, σ is the applied stress, ϕ is

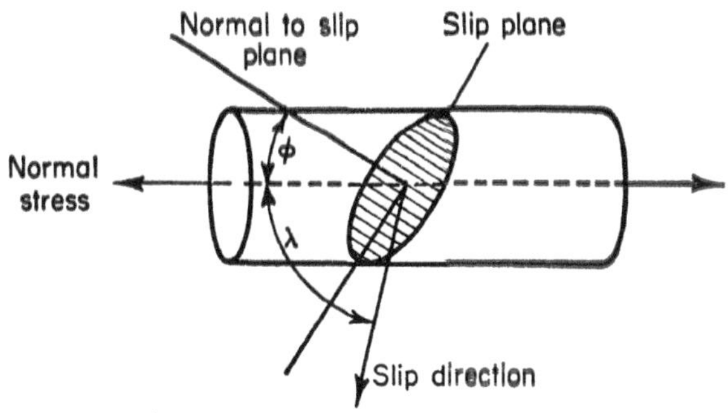

FIG. 7.5. Resolved shear stress in a cylindrical specimen.

the angle between the direction of the applied stress σ and the normal to the slip plane, and λ is the angle between σ and the slip direction. The value of τ is a constant for the material.

The resolved shear stress has a maximum value when the slip plane is at an angle of 45° to the tensile axis. If the slip plane is at an angle less than this ($\phi < 45°$) and so almost perpendicular to the tensile axis, then the application of a tensile stress will tend to separate the atoms by pulling them apart, rather than cause slip. Similarly, if the slip plane is almost parallel to the tensile axis ($\phi > 45°$) the value of σ is again small, since the area of slip plane involved is so large, and the stress tends to separate atoms rather than cause slip.

Individual crystals in a polycrystalline material will be oriented in different directions, and hence slip will commence first in those so oriented that the shear stress is high. The slip that occurs is then held up at the grain boundaries, and grain rotation occurs, which subsequently brings a further grain into the favourable orientation, and slip occurs in this one too. Such a process continues to propagate through a polycrystal, but some crystal grains will always be more favourably oriented and so show more slip, and contribute more to the overall deformation, than others. The smaller the grain size of the polycrystal, the greater the strength, since the more likely are the boundaries to cause interference between slip in adjacent crystals, owing to varying orientations.

The type of stress applied will affect the behaviour of a material, since it will vary the resolved shear stresses. Even a normally ductile material will not deform plastically if hydrostatic (triaxial) tension is applied, since the resolved shear stress on planes will be zero. It is impossible to apply triaxial tension in the strict sense, but conditions approaching this are believed to occur in certain circumstances, such as at the ends of very fine cracks, which may lead to brittle failure. Similarly, normally brittle materials may deform in a more plastic manner when the shear stresses are raised and normal stresses are kept low. Such conditions involve hydrostatic pressure, as in extrusion. By applying hydrostatic pressure around a material while under extrusion, considerable ductility has been demonstrated (by Pugh). The energy absorbed in overcoming the resistance to slip during deformation is largely converted into heat, but in the later stages of deformation the resistance increases and up to 15% of the energy remains locked up in the lattice as internal stress.

The plastic deformation of a material leads to changed properties in the crystal, and if strained in another direction the specimen shows a reduction in yield point—the *Bauschinger effect*. This arises from the unequal distribution of strain as a result of varying orientation through the grains. When the stress is removed, the internal stress resulting from the unequal distribution operates in a manner to assist the application of stress in another direction. The applied stress required to cause slip in this new direction is thus lower than the normal yield strength.

7.3.2. *The role of dislocations in plastic flow*

The shear stress criterion of the previous section was concerned only with the macroscopic effects. The dislocation density and the ease of travel of dislocations materially affect the quantitative nature of the plastic flow. Since an edge dislocation is reasonably symmetrical (Figure 4.4) the strain field around it is also symmetrical. When a shear stress operates along the slip plane, the dislocation is displaced, and this symmetry upset. The resistance produced by the disturbed atoms will increase the force required to drive the dislocation through the lattice, and this resistance is referred to as the *Peierls–Nabarro force*. The original conception of this force was based on the directional nature of bonds in homopolar crystals such as diamond and silicon, but it is now recognized as relevant to metals, particularly those with body-centred cubic lattices. The dissociation of a screw dislocation on several planes gives rise to a type of Peierls–Nabarro force, which explains in part the temperature-dependence of ductility in these metals. Similar arguments can be applied to the effect of stress and temperature on slip, in close-packed hexagonal metals, on planes other than the close-packed basal plane (0001).

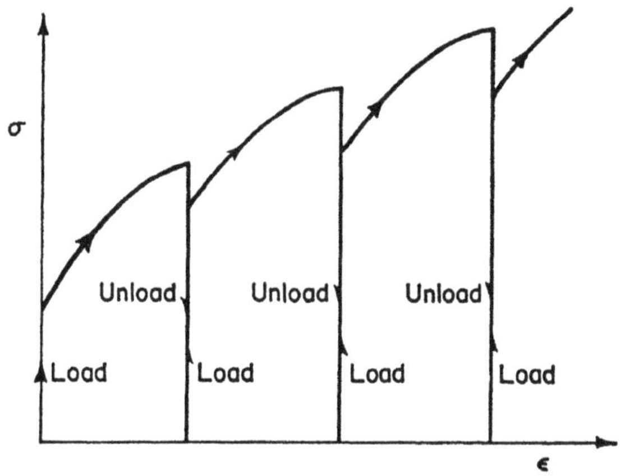

FIG. 7.6. Work-hardening by repeated loading.

The effect of grain boundaries on the slip process has been mentioned. Dislocations moving along slip planes cannot cross the boundary and continue to glide, because of the change in orientation. They pile up against the boundary, therefore, and eventually may assist the applied stress in starting slip in the adjacent grain.

Dislocations can interfere with each other, particularly when they are travelling on intersecting slip planes. The more slip systems that can exist in a particular structure, the more likely is this to happen. Cubic metals therefore show more *work-hardening*, as it is termed, than hexagonal metals. Work-hardening is defined as the increase in resistance to plastic flow once plastic flow commences at the yield point. The strength of the material is increased by work-hardening, but the ductility goes down (Figure 7.6). Electrical resistivity increases also, and density is lowered.

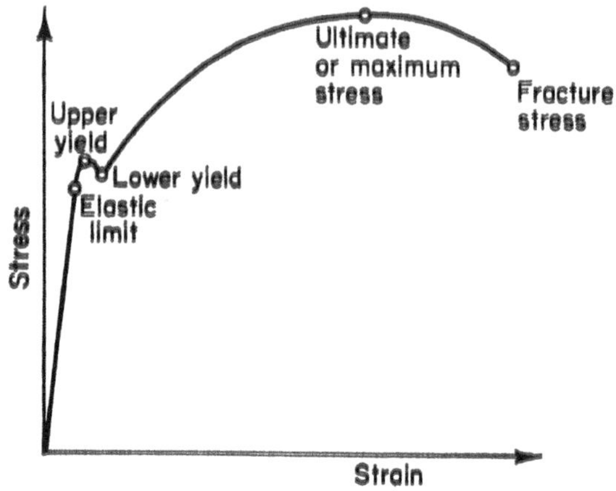

FIG. 7.7. Yielding in steel.

Solute or impurity atoms present in the lattice will also interact with dislocations. Figure 4.4 shows that the atoms above the slip plane are compressed by the presence of the dislocation, and those below are subject to tensile forces. Any solute atom larger than the matrix atom will therefore readily fit into the expanded region below the slip plane in Figure 4.4, and a smaller (substitutional) atom could relieve the stress by migrating to the upper, compressed regions. There is therefore a driving force to assist the segregation of solute atoms around dislocations. These segregations are called *Cottrell atmospheres* after the first man to describe them. When a dislocation is surrounded by such an atmosphere, the resistance to its movement is increased, since the solute atoms must be dragged along with it. The dislocation is said to be *pinned* by the solute atoms, and an extra stress is required to unpin it. This is one explanation of the sudden yielding, at higher stress, of materials such as carbon steel, where the carbon atoms form the Cottrell atmosphere. Once the dislocation has been dragged away from its atmosphere, the solute atoms have to diffuse along the slip plane to follow it, or to diffuse to another or following dislocation. The stress required to move the dislocation drops therefore so long as the atmosphere is separated from it, but later increases again, owing to work-hardening, or when the solute atoms catch up again, or affect new dislocations. There is therefore a lower yield point as well as an upper one, followed by the normal plastic flow relationship between stress and strain (Figure 7.7). Temperature has a pronounced effect on the diffusion of solute atoms, and in low-carbon steels there is a temperature range over which the solute atoms remain attached to the dislocation, or catch up with it so quickly that the atmosphere continuously raises the yield strength. This is the so-called *blue-brittle* range of temperature for steels. Above this range, the ease of movement of dislocations has increased so much that the atmosphere has little influence, and the steel becomes ductile again. Other materials also show the sharp yield point, and the existence of a drop following unpinning of the dislocations. COTTRELL confirmed the hypothesis by introducing nitrogen into zinc and artificially producing a sharp yield point as a result; the nitrogen atoms formed an atmosphere that inhibited the dislocation movement through the zinc lattice. Solute atoms do not have to form atmospheres around dislocations in order to impede their movement. Vacant lattice sites and interstitial atoms also influence the dislocation movement, sometimes by forming jogs (Figure 4.10), which must move with the dislocation.

When solute atoms cluster together, they form a barrier that the dislocation must either push through, or go round, in order to pass. Furthermore, a definite strain exists in the lattice as a result of the solute atoms, which inevitably increases the friction stress opposing

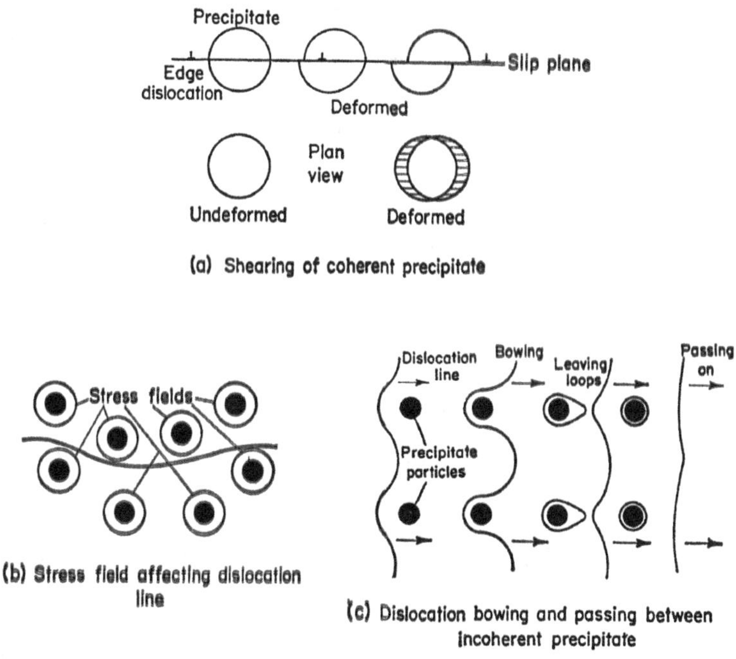

FIG. 7.8. (*a*) Shearing of coherent precipitate. (*b*) and (*c*) Dislocation movement affected by precipitates.

the dislocation movement. If the solute atoms form a precipitate, which remains *coherent* with the lattice (i.e. a definite orientation relationship exists on the slip planes or in slip directions), then the maximum strain and barrier effect is produced. If the precipitate grows in size so as to become *incoherent* and no longer related to the original lattice, then the strain decreases again. These effects are utilized in the precipitation hardening alloys such as duralumin. Dislocations may go through, go round, or break between, precipitates (Figure 7.8).

7.3.3. *The effect of temperature*

The loss of ductility as a result of work-hardening is important from many aspects, but particularly that of metal fabrication, where the required change in shape may not be achieved before work-hardening becomes so great that fracture will follow if the stress is continued. The density of dislocations increases during work-hardening, so that the strain energy may be very high. At this stage, the material may be softened by heating to a temperature greater than roughly one-third of its melting point (expressed in °K). The rise in temperature is necessary to activate those processes by which the lattice strains are relieved: diffusion, dislocation climb, and cross-slip. Nevertheless, time is required for the restoration processes to go to completion, and the higher the temperature the shorter the time.

The first process is called *recovery*. In Chapter 4 (page 98) mention was made of polygonization by climb of dislocations into cell-wall groupings. This produces recovery by virtue of rearrangement of dislocations, and since the movement of vacancies is involved in climb (Chapter 4, page 97) there are changes also in electrical resistivity. Recovery can also take place by annihilation of positive and negative dislocations, with or without the added movements in climb or cross-slip.

A second important softening process is that of *recrystallization.* The strained lattice can begin to form a new and unstrained one, given sufficient thermal activation, by a nucleation and growth process. The nuclei of the new grains are in the deformed matrix, and the amount of prior deformation has a considerable influence on both nucleation and growth. The greater the work-hardening, the lower the temperature at which nucleation of new grains commences, and the slower the growth at the lower temperature. In this case a very fine grain size results in the recrystallized material. The presence of solute atoms inhibits these recrystallization phenomena, and many very pure metals have a recrystallization temperature below room temperature (e.g. aluminium). Material of commercial purity may have a higher recrystallization temperature. Those materials whose recrystallization temperatures are below room temperature will be continuously softened when they are deformed at room temperature. Tin and lead are examples, and hence show great ductility. Quartz recrystallizes comparatively slowly at 500°C, but

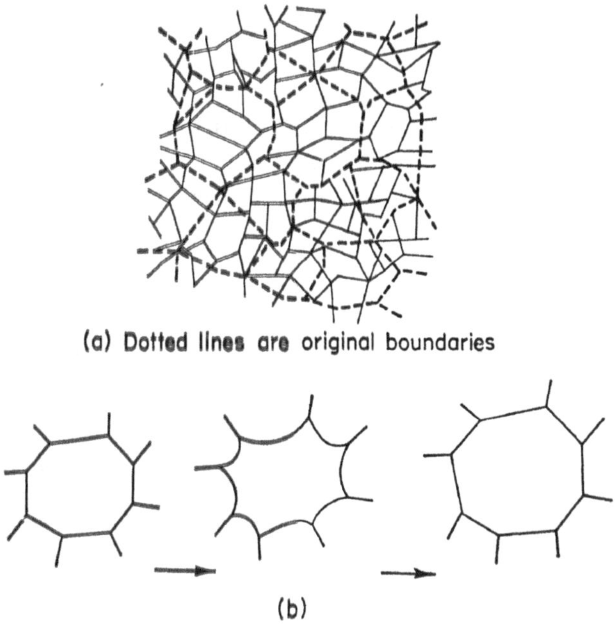

FIG. 7.9. (*a*) Stages in recrystallization. (*b*) Stages in grain growth.

recrystallizes in a few hours at 800°C, the rate roughly doubling for each 100 degC rise.

Once a material has recrystallized, further heating at the same temperature gives rise to further reduction in the total surface area of grains, and larger grains tend to grow at the expense of small ones. This is known as *grain growth* and results from long annealing times at high temperatures (Figure 7.9). By control of this process after straining, it is possible to grow single crystals in a large specimen, i.e. to arrange growth to eliminate all grain boundaries. In a commercial material, grain growth results in excessive softening, and is not desirable, except where softness is essential, as in metal seals or gaskets, where pressure is applied to force the metal into gaps between joints.

It follows, from the processes described above, that flow stress will decrease with rise in temperature, since the formation of dislocation jogs will be facilitated, and so will the diffusion of vacancies. The hardening effects of solute atoms will tend to decrease, since dislocations can by-pass them more readily. Similarly, precipitates will grow in size with rise in temperature, dislocations will face large precipitates in small well-defined areas, and will thus again find easy paths to avoid these barriers more effectively. In the precipitation-hardening materials, this state is known as the *over-aged* condition. The classic case of deformation at high temperature is, of course, that of creep, which will be dealt with in detail. The hot-working of metals and alloys (in which the material is deformed at high temperatures) is an example of utilizing the increased rates of deformation at higher temperatures for a given stress.

7.4. *Fracture*

When a ductile material is tested in tension, the plastic flow causes an elongation in the direction of the tensile stress, and a corresponding contraction or reduction of area. When the reduction of area becomes very large, the non-uniform distribution of appropriately oriented slip planes becomes evident, and in one (or occasionally more than one) particular spot a localized reduction commences, known as the *neck*. Deformation proceeds to some extent along the whole specimen from this point, but the sudden increase in stress that arises from the decrease in cross-sectional area inevitably causes more rapid flow around the neck. The stress continues to rise owing to the reduction in area and deformation will continue without increase in load.

At this point also, fracture commences in the centre of the test piece by the formation of cracks; these are transgranular, and grow in a plane perpendicular to the tensile axis. As the neck decreases the remaining area, the shear-stress distribution around the periphery causes rapid flow at 45° to the tensile axis, and the fracture changes from a rough plane to an inclined annulus, which finally meets the surface (Figure 7.10). This gives the characteristic *cup-and-cone* fracture associated with ductile materials. The fractured surfaces are not smooth, but fibrous, indicating the path of the plastic flow that finally caused failure. Face-centred cubic metals show this type of failure very well.

In more brittle materials, particularly the b.c.c. and c.p.h. metals, and the cubic ionic crystals, the fracture mechanism may be different, particularly at low temperatures. The characteristic brittle failure is by separation of crystals along well-defined planes, a *cleavage* failure. The fracture here exhibits bright facets of the crystal planes. Intercrystalline failure, where cracks form in grain boundaries and literally follow the actual boundary between grains, is often associated with brittle failure of a more complex nature, due either to phases in the grain boundary or to some extra agency such as a corrosive medium.

Ductile failure is not regarded as producing a crack in the normal sense of the word, i.e. a separation of atom planes with very high

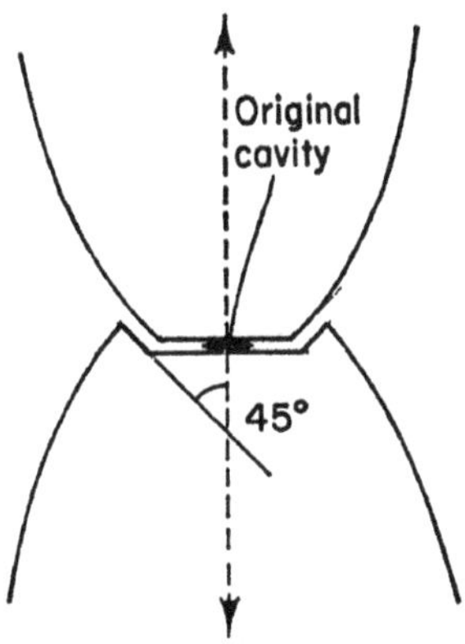

FIG. 7.10. Cup-and-cone ductile fracture.

concentration of stress at the edges. The creation of a cavity must be involved, this cavity growing outwards to the surface, where it meets the incoming boundary of the rapidly forming neck. In f.c.c. metals, the presence of impurity atoms has suggested that these are sites for nucleation of such cavities, and if no impurities were present the neck should grow inwards until failure occurs at a point. This is almost achieved with high-purity copper and aluminium. In b.c.c. metals, some of which do show ductile failure, the nucleation of cavities occurs at piled-up dislocation sides, e.g. at a grain boundary.

It is the brittle fracture that has recently received the most interest, particularly in view of the sudden and catastrophic failure of engineering components such as welded ships, pressure vessels in oil installations, and parts of electrical generators. Oxides and carbides, glasses, and some polymers also exhibit brittle fracture, and the modern theories to account for it apply in one form or another to all.

The earliest attempt to explain this was made by GRIFFITH in 1920, who proposed that glass possessed many fine cracks in the surface, which could propagate through the material and cause failure. The stress required to pull apart two adjacent layers of atoms depends on the interatomic distances and the surface energy of the faces after cleavage:

$$\sigma_f = \sqrt{(E\gamma/d)}$$

where E is Young's modulus, γ the surface energy, d the interatomic distance, and σ_f the fracture stress. The majority of materials do not approach this theoretical fracture stress closer than a factor of 10^{-2} E. Griffith imagined a crack of length a, at whose tip was a stress system able to cause parting of bonds one at a time, so that the applied stress could be very much lower than that required above. The work done in extending the crack remained as the surface energy γ of the adjacent faces produced. The same formula as that above is then applied with a different notation:

$$p = \sqrt{(E\gamma/a)}$$

where p is the smallest stress able to propagate a crack and a the crack length. These elastic cracks have been proved to exist in the surfaces of glasses and polymers, and the calculated stress required to propagate them is of the same order as the experimental values.

When first applied to metals, the Griffith approach required modification, for most metals exhibit some plastic flow before fracture, even around the advancing cracks produced at microscopic defects. OROWAN included in the Griffith formula a term to represent the plasticity of the material, but this cannot be verified owing to the difficulty of assessing the surface energy of the crack faces. Furthermore, the change from ductile to brittle behaviour in many metals and alloys as temperature is lowered cannot be accounted for by a formula in which no term is markedly dependent on temperature. The plastic behaviour that precedes fracture, however small in magnitude, must involve a slip mechanism, and the change from ductile to brittle behaviour over such a small temperature range must be due, at least in part, to some change in slip modes or interactions. If the yield stress at a particular temperature is less than the stress required to extend a Griffith crack, then ductile behaviour results. If the yield stress is larger, the material suffers brittle fracture. There is then, obviously, a critical range of temperature at which the values of yield stress and crack propagation stress cross over.

Dislocation mechanisms have been outlined by STROH to describe possible crack formations by pile-up of dislocations at grain boundaries. COTTRELL modified this to include coalescence of dislocations on intersecting slip planes (Figure 7.11). The latter type have been observed in MgO and TiO_2 crystals though not, as yet, in metals. These cracks, or rather dislocation cavities which then propagate as cracks, will increase in size only if the stress system allows this. If the dislocations can move in a normal way without coalescing, then the phenomenon need not begin at all. It is believed, however,

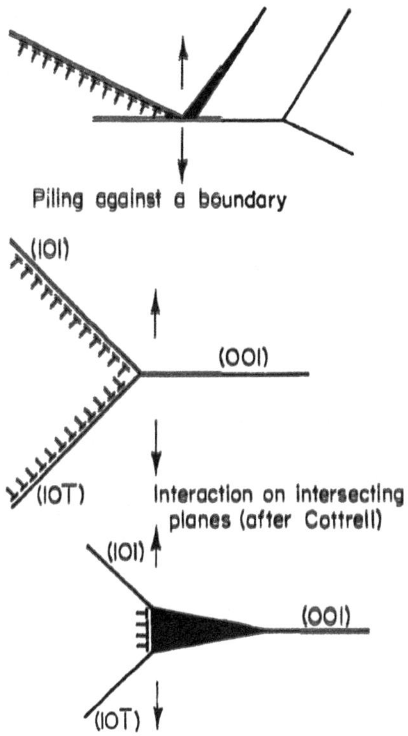

FIG. 7.11. Crack formation by dislocation interaction.

that nucleation of cracks can occur simply, and that it is the propagation stage that really governs the fracture mechanism. The yield stress therefore becomes a very important parameter, and attempts have been made to assess a theoretical value of yield stress in terms of (i) the friction stress required to drag dislocations through the lattice, (ii) the force required to drag dislocations away from the Cottrell atmospheres or other solute atom barriers, and (iii) the effect of piled-up dislocations at grain boundaries. This theoretical yield stress can then be compared with the crack-propagation stress calculated according to a Griffith–Orowan-type criterion. As yet, the results are not always in line with experiment, but they do approach quite closely the practical results on some materials.

An important criterion in this study is the transition temperature, below which fractures are brittle, and above which they are ductile. In components of large size made from steel, the requirements of site construction may involve low temperatures during the night (or day) and the danger of brittle fracture. Since many large pressure vessels and other units are welded on site (owing to their size), residual stresses may be created that are sufficiently great to cause fracture when the site temperature falls below the transition. To the residual stresses must be added the weight of components and supports, and restraining stresses due to supports, scaffolding, etc. This explains why some of the wartime 'liberty' ships fractured when lying in port, without operational loads, when the temperature fell below the transition and the residual stresses were sufficiently high.

To reduce this danger, welded structures can be stress-relieved, though often at great cost. The operational stresses may still exceed the fracture stress if the temperature falls, and the yield stress must therefore be kept lower than the stress to promote crack growth, or the transition temperature lowered. The means available for achieving these ends are: (i) low grain size (yield stress is inversely proportional to grain size); (ii) low impurities (less friction stress); and (iii) low strain rate (more chance for dislocation movement). Steels suitable for constructional work of this type are therefore deoxidized with aluminium to promote the formation of small grains, and contain a high ratio of manganese to carbon. Manganese helps to reduce grain size but, more important, it prevents carbon from increasing the friction stress. Nitrogen and hydrogen are avoided, for both cause increased brittleness, probably by lowering the crack-propagation stress. Silicon increases friction stress, and so must be kept low. Stress concentrations should be avoided, for the triaxial stresses that may result can increase the tendency to propagate cracks. Macroscopic defects, especially those due to welding, must also be avoided, and radiography of the welds is normally carried out to check on this point. Any such defects are normally cut out and rewelded.

In general, the likelihood of brittle fracture is increased by work-hardening because the friction stress increases as a result of interacting dislocations and the interference with slip mechanisms. Similarly, irradiation by neutrons can produce defects, that interfere with slip, and the friction stress or pile-up systems increase in value. The transition from ductile to brittle fracture occurs at a higher temperature as a result, and what was a normally ductile material may be rendered quite brittle. A change of +50 degC in transition temperature may occur in steels irradiated by a dose of 10^{19} neutrons cm^{-2}.

Metals may therefore display ductile or brittle behaviour, depending on the type of crystal lattice, the temperature, strain rate, and mode of stressing, in addition to the effect of composition and grain size. Ionic crystals and glasses normally behave in a brittle fashion, as indicated by the Griffith criterion. The preferred cleavage planes are usually those of close-packing, since these planes are generally farthest apart from each other. BRIDGMAN, as a result of high-pressure experiments, deduced that anisotropy of the strain tensor is mainly responsible for fracture. Sandstone to which pressure has been applied in one direction may break after removal of the load, with cleavage planes perpendicular to the direction of original pressure. The material suffered plastic flow when under pressure, and remained isotropic until the pressure was removed. At this point it became anisotropic, and so fractured. This is not unlike the Bauschinger effect that occurs in ductile materials.

7.5. *Creep*

When a constant stress is applied to a body over a period of time, plastic flow continues at a rate dependent on stress, time, and temperature, and the process is described generally as *creep*. Under conditions that give rise to creep, distinct stages of behaviour occur. Application of the load causes an instantaneous deformation ε_0, which is of significance only to a design engineer in allowances for initial clearances. The first stage of creep proper then begins, in which rapid deformation occurs, but at a decreasing rate with time, and hence is known as *transient*, *primary*, or *first-stage* creep.

At low temperatures and low stresses the transient flow follows a logarithmic law, obeying an equation of the form:

$$\varepsilon = A \log t$$

where ε is the strain, t the time, and A a constant. Pure f.c.c. metals and some c.p.h. metals behave in this way below room temperature, as also do some polymers and fibres such as glass.

At higher temperatures and stresses, the transient stage obeys a law of the form

$$\varepsilon = Bt^{1/3}$$

where B is another constant. This type of creep was first demonstrated by ANDRADE and is frequently named after him.

After the transient stage, the flow usually settles down into a *steady-state*, *quasi-viscous*, or *secondary* stage, which may be of exceedingly long duration. The equation of the curve over this period is of the form

$$\varepsilon = Kt$$

where K is another constant. The rate $\mathrm{d}\varepsilon/\mathrm{d}t$ varies markedly with temperature and/or stress, and can be represented by

$$\mathrm{d}\varepsilon/\mathrm{d}t = \text{constant} \exp(-Q/k\theta) \text{ for constant stress}$$

and $$\mathrm{d}\varepsilon/\mathrm{d}t = \text{constant} \exp(-\sigma/k\theta) \text{ for constant temperature}$$

where Q is the activation energy for steady-state creep, k is Boltzmann's constant, and θ the temperature. Subsequently, a rapid rise

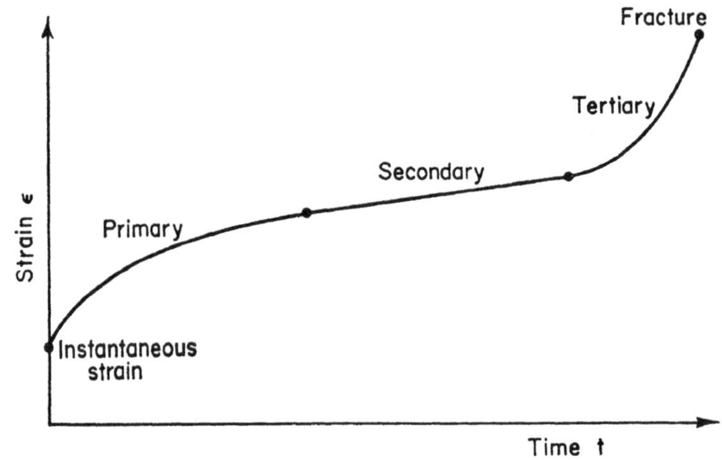

FIG. 7.12. Stages in the creep curve.

in the rate of deformation occurs again, and fracture follows. The accelerating, or *tertiary*, stage is associated with quite different mechanisms of deformation (Figure 7.12).

The present views on creep deformation are mostly related to dislocation movements by slip, the climb of dislocations associated with polygonization, and the sliding of grain boundaries. In the transient stage, dislocation movements are initially rapid, and even the barriers of solute atoms, etc., are readily overcome. As time proceeds, some barriers become more difficult to surmount, and new ones are created by interaction between dislocations. Thermal activation causes the dislocations to surmount the barriers, but at a decreasing rate, and hence the strain rate slows down.

In the steady state, the material has become partially work-hardened by virtue of the transient stage, i.e. interaction of dislocations has caused a number of pile-up situations near boundaries or barriers and/or the formation of tangled regions of complex dislocations. As deformation continues, so will the work-hardening. Since the temperature is high, however, the recovery processes are much more active also, and so cross-slip and the climb of edge dislocations will proceed to oppose the work-hardening. In the secondary creep stage, the work-hardening and recovery processes are assumed to balance out, giving a constant rate of deformation. The climb of edge dislocations, requiring the higher energy of activation, is believed to be the rate-controlling process, and hence the creep rate in this stage frequently increases exponentially with temperature. The formation of polygonized subgrains accounts for part of the recovery, and this can be followed by X-ray diffraction examination of suitably strained specimens.

During the steady state, a further process of deformation begins, by the sliding of grain boundaries, quite apart from slip within the grains. Such movements in boundaries have been shown by examination to involve the creation of vacancies, which then contribute to the process of dislocation climb, as described in Chapter 4 (page 96). Fine-grained materials have a greater surface area of boundaries, and hence sliding leads to a faster creep rate. Coarse-grained materials are thus preferred for creep-resistance. Diffusion processes, involving both solute atoms and vacancies, have a distinct effect on creep, and solute atoms showing rapid rates of diffusion in a matrix do not convey such a high creep-resistance to that matrix.

The tertiary stage of creep occurs largely as a result of the formation of microcracks in the grain boundaries, as discussed under fracture. Cavities are also believed to form in the boundary as a result of the migration of vacancies, possibly to sites nucleated by microcracking due to shear stresses and stress concentrations. The accelerating stage leads to fracture, frequently in a very short time. The formation of cavities caused by vacancy migration is most frequently associated with long periods of time at low stresses, leading to extensive recovery during a prolonged secondary stage. This is precisely the condition for which many materials are designed.

Apart from metals, ionic crystals show creep effects, but at very slow rates. In these, an ion must move over far greater distances than a metal atom during the slip process, in order to preserve electrical neutrality and finish up in the correct juxtaposition to the ions of opposite sign. The formation of polygonized structures, and the climb of dislocations, can occur in ionic crystals also, and hence favours deformation as against the probability of simple slip.

Creep-resistant materials can now be designed with a little more certainty than was attendant on the somewhat empirical methods of twenty or more years ago. The main requirements include the following:

(1) A base metal of close-packed crystal structure (f.c.c. and c.p.h.) and low stacking-fault energy (to give some difficulty in climb and cross-slip and hence lower recovery rates).
(2) Additions of solute atoms to harden the matrix, yet maintaining the essential features of (1) above (increasing the work-hardening without increasing recovery).
(3) Additions of alloying elements with solubility decreasing with decrease in temperature, to cause precipitation of coherent structure with the matrix, and a slow rate of diffusion at the service temperature (to increase resistance to slip by presenting barriers to dislocations).
(4) Additions of elements to form oxidation-resistant films on the surface (to prevent attack on a macroscopic scale which may reduce the load-bearing area).

7.6. *Fatigue*

Perhaps the best-known mode of failure, of metallic materials at least, is that of fatigue, in which an alternating stress well below the normal yield stress can produce rupture over a period of time. The design of components in this connection leaves much to be desired, and by far the greatest proportion of failures of engineering components in the past, if due to mechanical means, was as a result of bad design or bad workmanship. The essential requirement is to avoid stress concentrations, for otherwise the actual stress level is very much higher than is believed or designed for. Sharp radii, or even complete steps, in a tapering shaft or at changes of section must be absolutely rejected. Inspectors' stamp marks driven deep into an otherwise good metal surface are unnecessary and alternative markings can and must be found. Machining marks must be as fine as possible, and polishing to remove them can improve fatigue life enormously. Corrosive media can produce rapid changes in fatigue-resistance, either by the formation of etch pits that act as stress-raising points, or by changing the mechanism of crack formation and propagation, almost always accelerating it. More subtle design problems are involved where the stress cycle is complex, as in the flexing of aircraft wings or the strains in pressurized cabins, and these have led to exhaustive engineering tests to learn more about the mechanism of failure.

Metallurgical considerations are nevertheless important. It is well known that compressive stresses in the surface delay the formation of cracks there, and also that the strength of the metal is increased thereby, as a result of the inevitable work-hardening due to shot-peening, rolling, or polishing processes. Fatigue is now held to develop from a slip mechanism, the direction of slip changing during the stress cycle. Initially, work-hardening results uniformly over the whole structure, but very early in the fatigue life localized slip causes the appearance of coarse markings across the grains. Electron microscopy has shown that metal is actually extruded out of the slip bands in some cases, and intrusions occur in others. After further cycles of stress these markings begin to move laterally and cracks develop, which later spread along the slip bands; finally, rupture occurs.

It is now believed that the creation of point defects, such as are discussed in Chapter 4, is an important link in the fatigue mechanism. These defects interact with dislocations, and interfere with their normal movement. Such effects are more temperature-dependent than normal work-hardening, and this is true of fatigued specimens. The result of this interference with normal slip mechanisms is the lack of normal slip over a large portion of the grains, so that all further slip is localized in the coarse markings observed in the grains of fatigue specimens. Tests carried out at low temperatures where point defects are less mobile show less hardening, finer striations in the grains, and a rapid rise in electrical resistivity. Mechanisms have been suggested to account for the extrusion and intrusion on slip bands in terms of two different slip systems during the stress cycle (Cottrell) or cross-slip producing an internal cavity (Mott).

Recent work on ionic crystals, notably sodium chloride, silver chloride, and lithium fluoride, has indicated the existence of cross-slip in silver chloride but not in the other two, both of which have longer fatigue life than silver chloride. This would support the idea that cross-slip is a mechanism that affects fatigue failure. Fatigue, then, can be explained by the dislocation movements that take place as a result of alternating stress and the interference provided by vacancies and interstitial atoms. The long time required for propagation of cracks is reasonable when one remembers the low stress involved and that fatigue cracks often penetrate quite a large portion of the cross-sectional area before the final catastrophic rupture occurs.

7.7. *Thermal cycling*

Thermal expansion can take place uniformly along the three crystallographic axes of an isotropic body, but not necessarily in an anisotropic one. The non-uniform expansion that results from variations in interatomic distance in such materials can introduce plastic deformation between grains. Heating and cooling cycles thus lead to *growth* in the material, which is three-dimensional for a random orientation in a polycrystal, but much more directional in a polycrystal showing preferred orientation. Many c.p.h. metals show this effect, and other types of lattice may produce even worse effects. This growth due to thermal cycling has given rise to problems in nuclear reactor fuels employing alpha-uranium metal (orthorhombic), and has been cited as one cause of fatigue failure in bearing metals based on tin (tetragonal). Zirconium (c.p.h.) is another example, and its applications for nuclear reactor purposes may require further study. Some oxides and other compounds suffer a similar fate on thermal cycling.

8

Metals and Alloys

8.1. *Properties of pure metals*

The characteristics of the metallic lattice have been referred to in previous chapters and some comparison made of variations in particular properties between the metallic elements. Lists of those properties, of major interest in the selection of engineering materials, are included in Table III and Tables V–XI. Other properties, not included in these tables, are required for more specific applications, e.g. latent heats of fusion in casting and heat-transfer calculations; neutron capture and scattering cross-sections for nuclear engineering. Such data are under continuous revision as a result of improved experimental techniques and have not been included for this reason.

The purity of the element involved may seriously affect the values quoted in these tables. The electrical resistivity value of silicon and germanium, for example, has been undergoing continuous change as a result of improvements in zone-refining methods. Values should therefore be checked against the most recent information where great accuracy is essential. Other factors affecting properties will be referred to in Chapter 10.

The selection of metallic materials rarely leads to optimization on a pure metal, but most frequently suggests an alloy, or a combination of metal and compound, as in the cermets. The basic metal, however, still serves as a guide to the type of alloy or compound to be chosen. It is therefore necessary to see how the various types of alloy may be formed, and the inherent modification to the properties of the constituent metals that will result, as a guide to the selection of materials

TABLE VII

Thermal conductivities of metals and alloys of engineering interest

Material		Thermal conductivity (cal cm^{-2} degC^{-1} sec^{-1}) (at 200°C unless otherwise stated)	Material		Thermal conductivity (cal cm^{-2} degC^{-1} sec^{-1}) (at 200°C unless otherwise stated)
Silver	Ag	1·0 (0°C)	Iron	Fe	0·175
Copper	Cu	0·934	Platinum	Pt }	0·17
Gold	Au	0·71	Palladium	Pd }	
Aluminium	Al	0·53 (25°C)	Cobalt	Co	0·165
Duralumin (4% Cu alloy in Al)		0·40	Chromium	Cr }	0·16
			Tin	Sn }	
Magnesium	Mg	0·37	Germanium	Ge	0·14
Rhodium	Rh	0·36	Tantalum	Ta	0·13
Iridium	Ir	0·355	Niobium	Nb	0·125
Beryllium	Be	0·35	0·2% Carbon steel		0·12
Molybdenum	Mo	0·34	0·4% Carbon steel		0·115
Sodium	Na }	0·31	0·8% Carbon steel		0·110
Tungsten	W }		Lead	Pb	0·085
Brass, 70% Cu; 30% Zn		0·30	Vanadium	V	0·074
Zinc	Zn	0·27	Monel, 70% Ni; 30% Cu		0·06
Nickel	Ni	0·215	Zirconium	Zr }	0·05
Cadmium	Cd	0·22	Titanium	Ti }	
Osmium	Os	0·21	Antimony	Sb	0·045
Silicon	Si	0·20	Austenitic steel, 18% Cr; 8% Ni		0·035
Bronze, 90% Cu; 10% Sn		0·20	Bismuth	Bi	0·035

of this kind. Furthermore, the alloying of metals is very much concerned with the problems of reaction between two or more metals when both are in the solid state, or one in the solid and one in the liquid. (The author's first recollection of the term *compatibility* being used to describe this particular feature was that by K. Q. BAGLEY of the U.K.A.E.A., in 1952.) The presence of two dissimilar metals in close proximity rarely produces ill effects at low temperatures, but so many engineering applications now involve high temperatures, high pressures, and long times, that unintentional alloying during service may be a serious problem.

There are two methods of approach to the formation of alloys, one involving thermodynamic considerations, i.e. free energy; and the other involving mostly physical and mechanical effects, i.e. atomic

TABLE VIII

Coefficients of linear thermal expansion of metals and alloys of engineering interest

Material	Coefficient α (10^{-6} degC^{-1})	Range of measurement (°C)
Sodium	71	15–20
Selenium	37	25–45
†Cadmium	29·8	15–30
†Zinc	29·2	20–45
Lead	29·1	15–25
†Magnesium	25·8	15–25
†Indium	24·8	15–25
Aluminium	23·68	20–100
4% Cu duralumin alloy	21·6	15–25
†Tin	20·0	20–160
Brass, 70% Cu; 30% Zn	19·8	15–25
Silver	19·7	0–100
Bronze, 90% Cu; 10% Sn	18·0	15–25
Copper	16·6	15–25
Monel	14·4	15–25
Gold	14·2	15–25
†Bismuth	13·3	15–25
Nickel	13·3	20–30
†Beryllium	12·6	15–25
†Cobalt	12·3	20–50
Iron	11·7	15–25
0·2% Carbon steel	11·7	15–25
Palladium	11·67	15–25
0·4% Carbon steel	11·35	15–25
†Antimony	10·8	0–100
0·8% Carbon steel	10·8	15–25
Austenitic steel, 18% Cr; 8% Ni	9·0	0–100
Vanadium	8·95	15–25
Platinum	8·9	15–25
Cast irons	8·8–9·6	15–25
Rhodium	8·5	15–25
†Titanium	8·4	15–25
Niobium	7·1	15–25
Tantalum	6·6	20–30
Iridium	6·5	15–25
Chromium	6·2	15–25
Germanium	6·0	0–100
†Zirconium	5·89	15–25
Molybdenum	5·44	15–25
Tungsten	4·45	0–500
Silicon	4·68	15–25
Osmium	4·60	40–60

† Mean figures for a polycrystal of anisotropic material

TABLE IX

Specific heats of some metals and alloys of engineering interest

Material	Specific heat (cal g^{-1} degC^{-1})	Material	Specific heat (cal g^{-1} degC^{-1})
Beryllium	0·425	Selenium	0·081
Sodium	0·295	Molybdenum	0·066
Magnesium	0·250	Zirconium	0·0659
Aluminium	0·214	Niobium	0·0642
Silicon	0·1597	Silver	0·056
Titanium	0·125	Tin	0·0542
Vanadium	0·120	Antimony	0·0504
Iron	0·107	Tantalum	0·034
Chromium	0·1068	Mercury	0·033
Cobalt	0·1056	Tungsten	0·032
Nickel	0·105	Platinum	0·0314
Zinc	0·0925	Gold	0·0312
Copper	0·092	Lead	0·0310
Germanium	0·086	Uranium	0·028

size and lattice strain. The thermodynamic approach is rapidly gaining ground as methods of measurement improve, but the small energy changes involved in many of these processes create difficulties in quantitative assessment. The physical basis is better established, since X-ray diffraction measurements, of high sensitivity, have been available for over thirty years. (Reference should be made to metallurgical textbooks for the thermodynamic treatment.) The second method allows of classification into broad groups of behaviour, and will be used here. Only binary alloys will be given as examples.

8.2. *Compatibility*

Before examining the types of alloy formed, the processes by which compatibility may be affected, other than the mutual solubility of metals, must be examined.

Although two dissimilar metals may be mutually soluble in the liquid state, and even solidify as a single solid solution, they can react together in the solid state only by the process of *diffusion*. As the name implies, diffusion involves the relative movement of atoms, and is of importance not only in compatibility but in transformations in metallic alloys in the solid state, in the sintering of metal powders, and in oxidation and creep at elevated temperatures.

TABLE X

Atomic radii (closest distance of approach) and densities of elements of likely interest in engineering materials

Element		Atomic radius (ångström units)	Density (g ml^{-1})
Caesium	Cs	2·62	1·95
Rubidium	Rb	2·485	1·53
Potassium	K	2·31	0·88
Sodium	Na	1·855	0·97
Lead	Pb	1·745	11·3
Magnesium	Mg	1·60	1·74
Zirconium	Zr	1·59	6·4
Bismuth	Bi	1·56	9·8
Tin	Sn	1·508	7·3
Cadmium	Cd	1·486	8·65
Titanium	Ti	1·46	4·51
Antimony	Sb	1·45	6·67
Silver	Ag	1·44	10·5
Gold	Au	1·435	19·3
Niobium	Nb	1·43	8·6
Aluminium	Al	1·43	2·7
Tantalum	Ta	1·425	16·6
Tungsten	W	1·41	19·3
Uranium	U	1·38	19·05
Platinum	Pt	1·38	21·45
Molybdenum	Mo	1·36	10·2
Zinc	Zn	1·33	7·14
Vanadium	V	1·313	6·15
Copper	Cu	1·275	8·9
Iron	Fe	1·26	7·9
Cobalt	Co	1·25	8·7
Nickel	Ni	1·25	8·9
Arsenic	As	1·25	5·7
Chromium	Cr	1·246	7·1
Germanium	Ge	1·22	5·4
Silicon	Si	1·175	2·34
Selenium	Se	1·16	4·85
Manganese	Mn	1·13	7·4
Beryllium	Be	1·125	1·85
Phosphorus	P	1·08 (red)	1·82(white)
Sulphur	S	1·06	2·05
Boron	B	0·97	2·3
Carbon (graphite)	C	0·77	2·3
Oxygen	O	0·6	1·13 (−184°C)
Nitrogen	N	0·53	0·81 (−195°C)
Hydrogen	H	0·37	0·07 (−252°C)

TABLE XI

Moduli of elasticity of some metals and alloys of interest in engineering materials

Material	Modulus of elasticity (10^6 lb in^{-2})	Material	Modulus of elasticity (10^6 lb in^{-2})
Osmium	80	Copper	17
Iridium	74	Palladium	
Rhenium	67	70/30 brass	16
Boron	64	90/10 bronze	
Tungsten	59	Silicon	15·5
Molybdenum	47	Titanium	
Beryllium	44	Niobium	15·2
Chromium	42	Zirconium	13·7
Rhodium	41	Zinc	12
Iron		Antimony	11·3
Nickel	30	Silver	11
Cobalt		Gold	10·8
Carbon steels	30	Aluminium	10
Austenitic steels		Selenium	8·4
Uranium	27	Cadmium	8
Tantalum		Magnesium	6·5
Monel alloy	26	Tin	5·9
Platinum	22	Bismuth	4·6
Hafnium	20	Lead	2·6
Vanadium	19		

The simplest process of diffusion arises from a concentration gradient. If two metals A and B are brought into intimate contact, the concentration of A on one side of the interface is 100%, and on the other, zero; likewise, but on the opposite side, for B. There exists, therefore, an opportunity for A and B atoms to move across the interface to even out the concentration of both. This occurs first by interpenetration of the two surfaces (by no means smooth—see Figure 4.1) and then by one or more of several mechanisms described below. The relationship between the two transports forms the basis of FICK's law of diffusion:

$$dN/d\,t = -D\,dC/dx \qquad (8.1)$$

where N is the number of atoms per unit area diffusing in time t, C the concentration of one atom in distance x, and D is the diffusion coefficient. The diffusion coefficient depends on concentration and on temperature. The higher the temperature and concentration gradient, the more rapid the diffusion.

When an atom of the second metal B is small enough to occupy voids in the crystal lattice of A, i.e. to become *interstitial* (see Chapter 3), then movement can take place with the least disturbance to the lattice of A, and so may be rapid. Only a few atoms are small enough to satisfy this condition, e.g. carbon, nitrogen, and hydrogen in the lattice of iron.

The possible occupation of a lattice site by the second atom, i.e. a *substitutional* position, is relatively easy to assess, and is governed by conditions described later. The movements through the lattice cannot take place, however, without the relative change of position of both types of atom, and this would not be achieved but for the presence of vacant lattice sites. Where a vacancy occurs an atom of

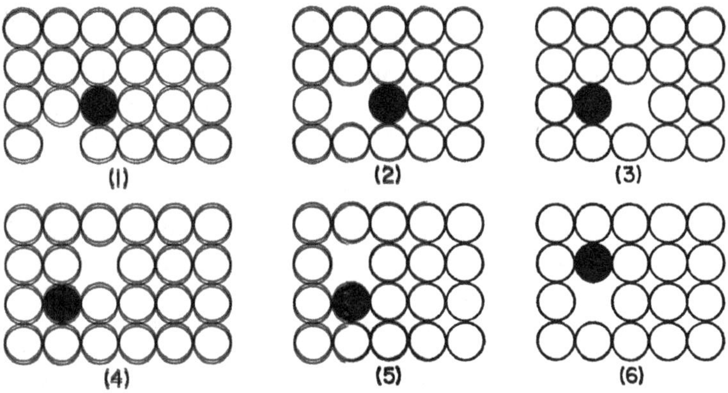

FIG. 8.1. Solute atom–vacancy movements in diffusion.

the parent metal A can move into it, then an atom of B can occupy the former A position, and so on. Figure 8.1 shows the relative movements involved. This type of transport is known as *vacancy diffusion*. It is complicated by the fact that the vacancy–solute atom pair tends to remain stable, and although the vacancy itself may move, it generally does so only in the vicinity of the solute atom. This assists diffusion of the solute atom, since it retains a vacancy to help in the reshuffle of relative positions of A and B atoms, but it ties up a vacant lattice site so that this is not available to other atoms. The concentration of vacancies and the relative stability of the vacancy–solute atom pair obviously affect the rate of diffusion in this process.

Diffusion occurs between like atoms, even in a pure metal, by mere change of relative positions in the lattice. This can occur only by vacancy diffusion, but since there is no solute atom to form a pair, the position is less complex.

In the diffusion of two dissimilar metals A and B it must not be assumed that both diffuse into each other at the same rate. There are two diffusion coefficients involved, D_A and D_B, and they are rarely equal. This was first demonstrated by KIRKENDALL. Since the diffusion of the faster atom is not compensated by substitution of as many atoms of the slower one there must be a greater concentration of vacancies on the side of the interface from which the faster-moving atom originates. Porosity may develop as a result.

Diffusion occurs more rapidly in the vicinity of grain boundaries and of dislocations, owing to the irregular lattice and the obvious occurrence of vacancies. A lattice under strain assists the diffusion process, but external pressure does not have much effect. (The initial pressure between two dissimilar surfaces merely ensures good contact.) Close-packed lattices decrease the rate of diffusion, more open ones increase it.

When a solid metal A is in contact with a liquid metal B the interface may be very similar to that in the previous case of solid/solid reaction. The liquid B can be regarded as a condensed layer only a few atoms thick on the surface of A. Diffusion will then occur exactly as described above. The supply of B atoms from the liquid, to replace those already diffused into A, will be simple, and because of the increased energy of the liquid it will occur readily. Since the temperature will be high (unless B is a metal of very low melting point) the rate of diffusion will be high. In the case of solid/liquid interaction, the irregularity of the grain boundary of the solid becomes important, and diffusion along these paths is frequently so rapid as to cause grains to be entirely separated from the solid and to fall out into the liquid, where they proceed to dissolve by diffusion.

Liquid/liquid reaction will, of course, be extremely rapid, the diffusion mechanism being much simpler in the liquid state, with its short-range order. If the two metals are not miscible in the liquid state, however, they will separate into two layers, as with lead and zinc.

Compatibility of metals thus depends on two main factors, the type of alloy system that can exist between the constituents (which represents the equilibrium state after long times of contact), and the rates of diffusion of the constituents in the presence of each other.

8.3. *Interstitial solid solution*

As previously mentioned, a solute atom can occupy one of the voids in a crystal lattice if it is small enough. Only hydrogen, boron, carbon, and nitrogen (occasionally oxygen) can do this, and in some solvent metals even these atoms are too large to fit. If the lattice of the solvent can be strained elastically they may still fit into these interstitial positions, but not in great quantity, for the strain in the lattice would eventually cause instability. Carbon in the b.c.c. form of iron will dissolve only to the limit of 0·02% by weight (% w/w), since the diameter of the carbon atom exceeds by more than a factor of 2 the size of the interstitial void at a tetrahedral site. However, the carbon atom actually occupies the even smaller octahedral site, probably because the iron atoms at the cube edges are more easily stretched to accommodate it. In f.c.c. iron, the carbon solubility is 1·7% w/w, although the octahedral site is still somewhat smaller than the diameter of the carbon atom. Nitrogen, similarly, dissolves interstitially in f.c.c. iron, the solubility being greater than for carbon (2·8% w/w), as the nitrogen atom is the smaller of the two.

The largest available sites for interstitial atoms in metallic lattices were discussed in Chapter 3. For octahedral sites in f.c.c. lattices the site is $0{\cdot}41R$, where R is the radius of the solvent atom. True interstitial solid solution will exist for these sites only when b/a is equal to or not much greater than 0·41, where a and b are the atomic radii of solvent and solute respectively.

The straining of the lattice to accommodate larger atoms in interstitial positions is feasible, as previously suggested, but in fact the solute atoms often take up specific lattice sites in relation to a number of solvent atoms, and are said to form an interstitial compound. Many hydrides, borides, carbides, and nitrides are formed by the transition metals as a result of this stretching of the lattice. The simpler interstitial compounds exist when $0{\cdot}41 < b/a < 0{\cdot}59$. The compounds formed take the formulae AB (sometimes) A_2BAB. In the composition, the NaCl structure is adopted, in which each interstitial atom has six neighbours. This is typified by the carbides and nitrides of titanium, zirconium, vanadium, niobium, and tantalum, and some of their hydrides.

The simple structure of these interstitial compounds cannot be maintained when b/a is greater than 0·59, but in certain cases the lattice can still stretch a little farther, as with carbon in alpha-iron, forming Fe_3C *cementite*, for which $b/a = 0{\cdot}63$. The structure is then complex. Similar examples include the carbides of chromium and cobalt, and some of the borides.

A third set of compounds exists, between metal atoms, in which the variation in atomic size is about 30%. These are the *Laves* phases, an arrangement based on the formula AB_2. Each A atom has 12 neighbours of atom B and 4 of its own. Each B atom has six of A and six of B as near neighbours. Three closely related structures illustrate the typical arrangement of these phases, those of $MgCu_2$ (cubic), $MgNi_2$ (hexagonal), and $MgZn_2$ (hexagonal). Similar compounds occur in many alloy systems, but as yet have no great commercial or engineering significance.

8.4. *Substitutional solid solution*

Atoms that are perfectly miscible in the liquid state do not necessarily remain so in the solid state. They may be only partially soluble (over a restricted range of temperature or composition), or be completely insoluble.

Where complete or partial solid solubility exists and the solute atoms are too large to occupy interstitial positions, they replace solvent atoms at random in the lattice, and are accordingly referred to as *substitutional*. A solute atom of very large (or very small) radius cannot occupy such a position without straining the lattice, and hence there must be a limit to this strain before instability sets in. This is the reason for partial solubility over a restricted range of composition. As with interstitial solutions and compounds, empirical rules, first postulated by HUME-ROTHERY, can be applied to predict the probability of solid-solution formation.

If the two metals A and B have the same crystal lattice type, and their atomic diameters are a and b respectively, then for primary substitutional solid solution of B in A, over a substantial range of composition,

$$\frac{a-b}{a} < 0{\cdot}15$$

If A and B are of the same crystal type, but

$$\frac{a-b}{a} > 0{\cdot}15$$

then extensive solid solution is unlikely.

If A and B have different crystal lattices, then even if

$$\frac{a-b}{a} < 0{\cdot}15$$

extensive solid solutions are not likely to form.

The distortion of the lattice A by the introduction of B atoms can be minimized by rearrangement of the random substitution of B for A by a more regular one. This leads to the formation of a *superlattice* in which two superimposed lattic arrangements are set up, and the like atoms approach each other as closely as the regular pattern of the lattice will permit. Slow cooling produces this *ordered* arrangement, and the change is often referred to as an *order–disorder transformation.* Figure 8.2 shows one change in the lattice which might result.

When the atomic diameters a and b are nearly equal, there is no advantage in the formation of a superlattice, since the distortion in the disordered state is negligible. When a and b differ, but are still within the restrictions of primary solid solution $(a - b)/a < 0{\cdot}15$, then a superlattice is likely to form, to relieve the distortion. When

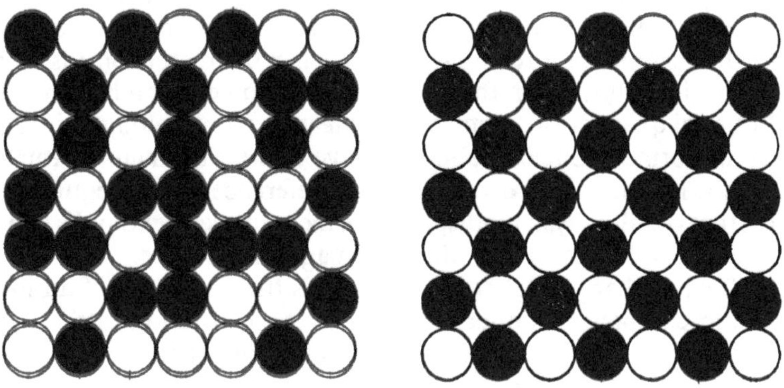

FIG. 8.2. Formation of a superlattice by ordering.

$(a - b)/a > 0{\cdot}15$, and solid solution is restricted, then lattice distortion is so great that an ordered lattice cannot much improve the situation, and a superlattice is unlikely.

At low concentrations of solute B the ordered lattice does not extend over the whole volume of alloy, since there are insufficient numbers of B atoms to do this. In this case, *domains* of order can be formed, among the normal or disordered lattice. Domains form and grow during normal superlattice formation, according to composition, temperature, and time of annealing.

Examples of substitutional solid solution are numerous. In the alloys of copper, silver, and gold (on which Hume-Rothery first worked) there is ample evidence of the influence of size-factor. Gold and silver have atoms of almost identical size and form solid-solution alloys in which the composition changes smoothly from pure silver to pure gold, without any interruption in the lattice. In copper, however, both gold and silver are on the borderline of favourable size; as a result gold forms a series of solid solutions, with superlattices at lower temperatures, and silver shows only restricted solid solution (8% w/w Ag in Cu, and 9% w/w Cu in Ag).

Copper also forms continuous solid solutions with nickel, the size of their atoms being almost identical; $(a - b)/a = 0{\cdot}025$. For copper and silver the ratio is 0·13; for gold and silver it is almost zero.

Many alloys of commercial importance arise from the formation of these solid solutions. Except near the limit of solubility (if one exists) the alloys are stable, and have many desirable properties (see below). As the limit of solubility approaches, where size-factor restriction exists, the lattice distortion grows to the point where the existing crystal lattice can no longer tolerate any increase in the amount of solute. Either the lattice changes to a different type, which can then support more solute by relieving the distortion (as in the copper-zinc alloys), or the two metals become insoluble with the additional quantity of solute, and a *eutectic* or low melting point alloy may be formed. Eutectic alloys behave like an intimate mixture of two phases on a microscopic scale. Reference to the equilibrium structures of alloy systems displayed in metallurgical textbooks should be made for details of this kind (see page 244).

A further change that can take place is the formation of an intermetallic compound, rather than a new solid-solution lattice, or a eutectic, though such compounds may be insoluble in the primary or other solid solutions and form a eutectic themselves. These compounds may be of several different types.

8.5. *Intermetallic compounds*

Reference has already been made to interstitial compounds and the Laves phases. For those metals in which primary solid solution may be restricted, there is a second effect that can alter the alloying propensity—that of their electrochemical nature. The degree of electronegativity of one metal to the other increases the tendency to form stable compounds. The electronegativity of elements in the periodic table increases from left to right through the periods, and from bottom to top in the groups (Table II). Thus, elements in the top right-hand corner of Table II are strongly electronegative (oxygen, fluorine, sulphur, chlorine). Those in the bottom left-hand corner are strongly electropositive (barium, caesium, cerium, thorium). Occasionally, some true valency compounds are formed, as with the strongly electropositive magnesium. Magnesium forms AB-type compounds with selenium (as do some other metals), AB_2-type compounds with silicon and tin, and A_3B_2 compounds with antimony arsenic, etc.

(These structures were covered in Chapter 3, referring to the NaCl, fluorite, and oxide-type lattices, respectively.) Here, the second metal is strongly electronegative to magnesium, but in order of electronegativity tends to form compounds of different stability or melting point. Mg_2Si melts at 1,085°C, whereas Mg_2Sn (with the same lattice) melts at 778°C, since silicon is more electronegative than tin.

Hume-Rothery drew attention to compounds in which the ratio of electrons to atoms was an important criterion in formation of the particular phase and its structure. The atomic diameter is believed to play some part also, and there are exceptions to this. The range of composition of *electron compounds*, as they are called, is often quite wide, probably by virtue of solid solution of one metal in the compound. Valency and degree of electronegativity also play some part.

The distinctive features of these compounds are as follows:

Electron/atom ratio 3/2, generally b.c.c. lattice, typified in the β phase of Cu–Zn alloys

Electron/atom ratio 21/13, generally complex b.c.c. lattice, typified in the γ phase of Cu–Zn alloys

Electron/atom ratio 7/4, generally c.p.h. lattice, typified in the ε phase of Cu–Zn alloys.

The γ and ε phases are also found in Cu–Sn and Cu–Al. The β phase occurs in Cu–Al, Al–Co, and Al–Ni alloys.

Electron compounds, like many others, tend to be stable structures, although some decompose before reaching the melting point. They tend to be hard and brittle, undergoing little plastic deformation. It is necessary to study the alloy system in which they occur as a whole, however, since the presence of these compounds is by no means necessarily detrimental to their applications, and in certain cases they are highly desirable as constituents of the system.

8.6. *Equilibrium and non-equilibrium*

The compatibility problem represents a case of non-equilibrium. The study of alloy systems, to identify the various phases present and their boundaries of composition and temperature, represents equilibrium. In practice, many problems of materials involve a state of non-equilibrium rather than the stable one, but inevitably, given sufficient time, equilibrium is approached. When temperatures are low, equilibrium may not be attained in the normal service lifetime of a component, and hence a metastable condition (if it exhibits the desired properties) can be acceptable. Any change in the conditions of service, however, may upset the metastability, and the approach to equilibrium may be more rapid, or even attained completely. A typical example is given by the heat treatment of alloys, such as steel. The quenching of a medium carbon steel (say 0·4% carbon) from a temperature of about 850°C into water produces a metastable hard and brittle structure called *martensite*. This structure is too brittle for most applications and is usually tempered, i.e. reheated to approach equilibrium more closely to modify the properties. The resultant structure produced by tempering at 300°C will not be the normal equilibrium state of the same steel slowly cooled from 850°C, but it will have greater strength and reasonable ductility. If, in service, the steel is at any time heated to a higher temperature than 300°C, it will again tend to equilibrium, and will lose some of the strength of the metastable state, while improving in ductility. If it should subsequently be heated to a temperature as high as 550–600°C for a reasonable period of time, there will be further changes in structure that will soften the steel even more than air-cooling from 850°C would have done in the first place. Its mechanical properties may therefore be completely ruined.

It is essential when considering an alloy to remember that the equilibrium diagram represents a special case, that of the ultimate state after allowing adequate time for diffusion and phase changes in the system. The majority of metallic alloys are not used in the equilibrium state, and the optimization of properties is based on the assumption that the service conditions are known precisely, or a reasonable estimate of the possible departure from them can be made. The metallurgist is frequently blamed for failures that strictly result from inadequate knowledge of the true service environment or failure to recognize the possibility of departure from design conditions. These effects are not of course restricted to metallic materials, but they require special emphasis in this case, in view of the extensive use of alloys in a metastable condition and the strong influence of temperature, time, and stress.

8.7. *The properties of alloys*

The properties of constituent metals in an alloy system are obviously considerably modified by the formation of solid solutions or compounds. Reference was made to the effect of impurity atoms in exploring the properties of materials, but the different phases formed between two metals in a binary alloy may add to the obvious effects of merely adding the solute atom.

The density changes when one metal is added to another, simply because the mass of the two atoms will not be the same. As the packing of the atoms is also affected, and dependent on whether a solid solution, eutectic, or compound results, then a second influence arises from structure. VEGARD'S law applies to some solid solutions, and assumes that the atomic concentration of solute produces a linear change in the size of the crystal lattice. For interstitial solid solution,

$$\rho_I = (n_A A + n_B B)m/V$$

where ρ is the density, n_A the number of atoms of metal A, A the mass of atom A, n_B the number of B atoms, B the mass of each B atom,

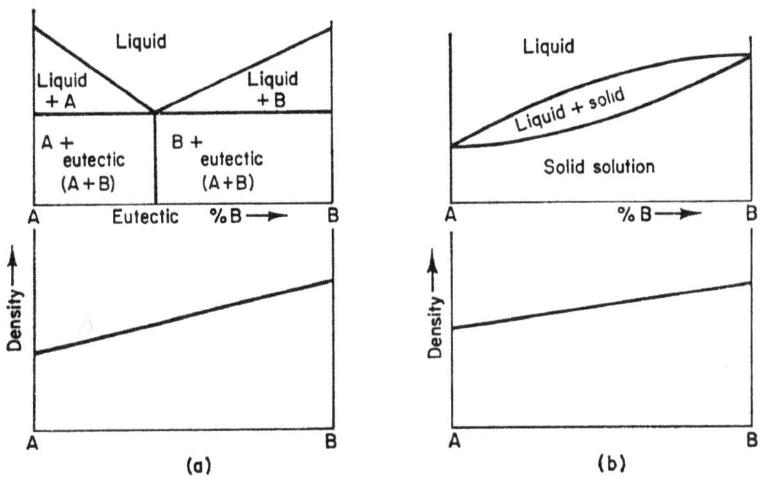

FIG. 8.3. Density of alloys in: (*a*) Simple eutectic system. (*b*) Solid-solution system.

m the mass of the hydrogen atom, and V the volume of unit cell. For substitutional solid solutions,

$$\rho_S = nM_A m/V$$

where n is the number of atoms and M_A average mass of A and B atoms. Figure 8.3 indicates this for a eutectic and a complete solid-solution system.

The mechanical strength of alloys depends largely on the type of alloy system. Eutectics behave as their structure would suggest, a combination of discrete areas of the constituent metals, with average properties based on the composition. Hardness changes in a binary eutectic in a simple linear fashion, a straight line joining the hardness figures for the individual metals or the terminal solid solutions. The plastic behaviour in tension or shear depends on the flow properties of the individual metals. A eutectic mixture of two soft metals, such as lead and tin, deforms much as a mixture would suggest. A hard constituent forming a eutectic with a softer one would behave more in line with the properties of the softer constituent. Binary eutectics are commonly formed between two solid solutions, and then the properties of these solid solutions are involved. Solid solutions are invariably stronger than the pure constituent forming the solvent, owing to the extra strain in the lattice and the interference with dislocation movement. Complete solid solubility leads to a maximum in the curve of strength against composition, where the hardening effect of each on the other reaches a limit. Binary systems with partial solid solubility and a eutectic between the two solutions combine these two basic systems. In all these arguments, the size of the dispersion of one constituent in the other is important. Even for complete solid solubility, the formation of a superlattice in the ordered state plays a part. Figure 8.4 shows typical results for hardness in systems such as those described above.

Conductivity, both thermal and electrical, is obviously affected by alloying. Again, for a eutectic, the mixture rule applies, and each area

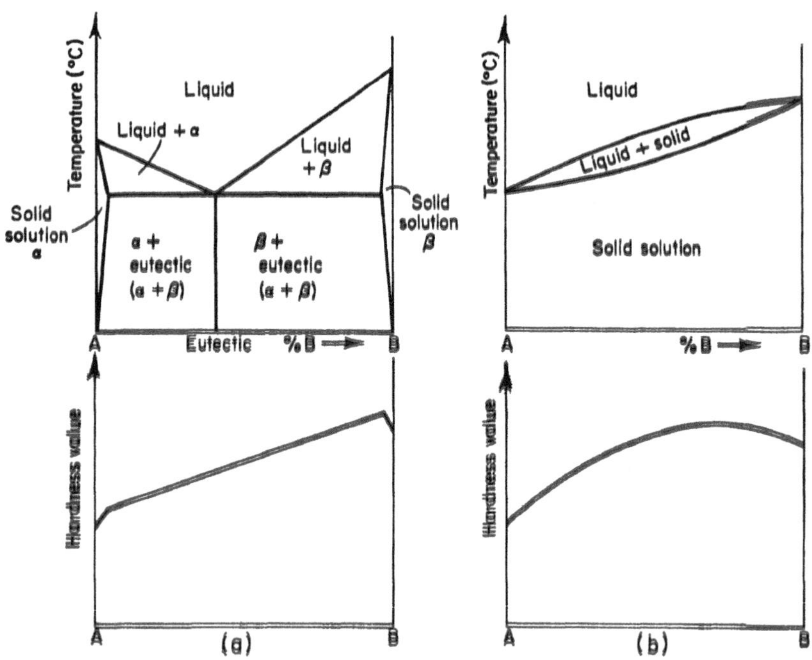

FIG. 8.4. Hardness values of: (*a*) Eutectic system with partial solubility. (*b*) System with complete solid solubility.

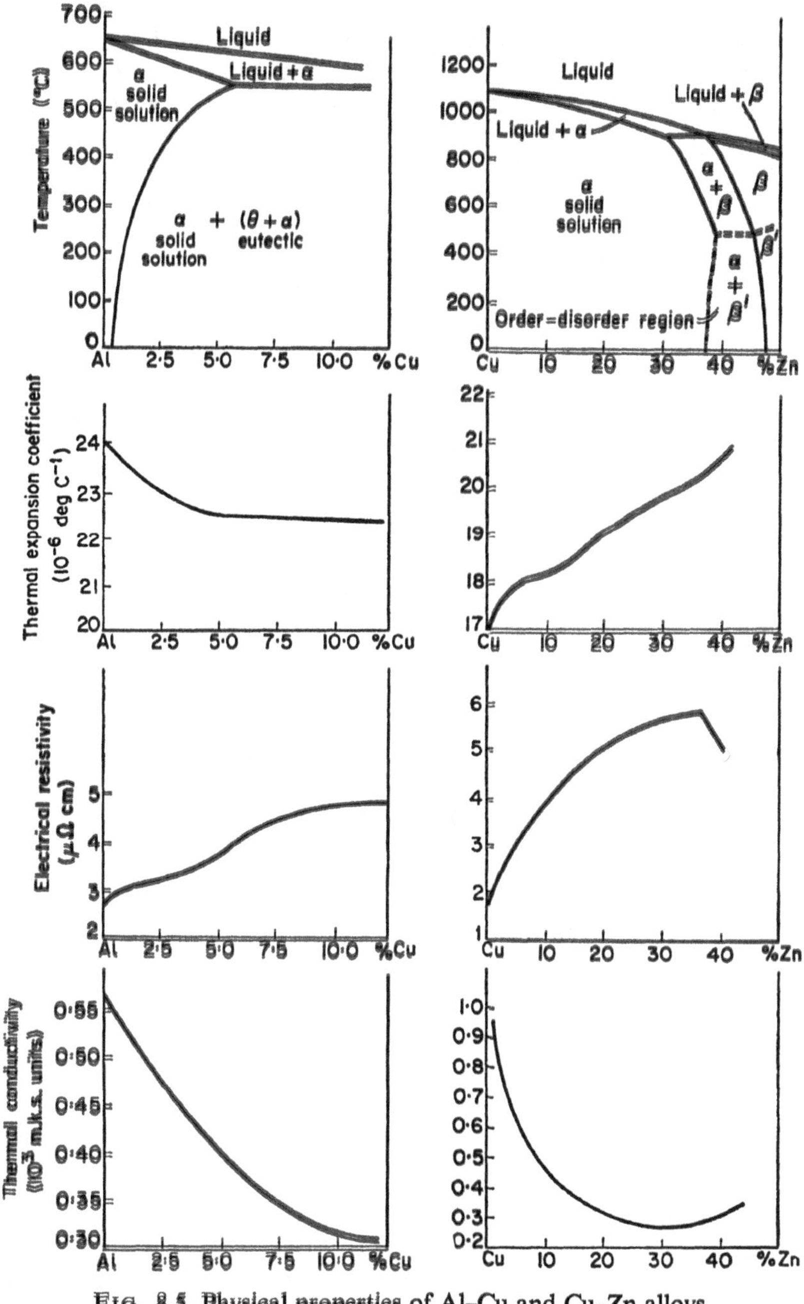

FIG. 8.5. Physical properties of Al–Cu and Cu–Zn alloys.

of each constituent makes a contribution, so that an average is obtained over the range of composition, according to the proportion of pure metal or solid solution and eutectic. In alloys showing solid solubility, the electrical conductivity must be affected by the irregularity introduced into the lattice and the new electron concentration around the atoms. Electrical conductivity always falls, therefore, from the value for the pure metal when a solute is added. Alloys with complete solid solubility hence show a minimum, often at a composition coincident with the maximum in the hardness curve. The addition of any soluble alloying element therefore invariably lowers the electrical conductivity. The effect on thermal properties will depend on the alloy system. Since the lattice contributes more to thermal than it does to electrical conductivity, the variation in lattice parameter as the proportion of solute increases must affect specific heat, thermal expansion, and thermal conductivity. In most alloy systems, conductivity tends to decrease, and specific heat tends to increase, in the formation of a solid solution. Thermal expansion normally varies in a manner not dissimilar to that of density, but if marked differences in cohesion occur as a result of alloying, then thermal expansion may be reduced. Figure 8.5 gives some examples of such property changes in solid-solution alloys.

If a superlattice is formed in a solid solution then the specific heat rises, since extra energy is necessary to form the superlattice. The ordered arrangement presents the ideal condition for conductivity, and hence both electrical and thermal conductivity values increase in the transition from disorder to order in such an alloy. The mechanical strength normally increases in the ordered state, but this depends very much on the size of the ordered domains, and there can be a softening when dislocation movement is assisted by the ordered state.

8.8. *Precipitation in alloys*

Some alloys exhibit a change in solid solubility with temperature, which can be utilized in the production of a metastable structure with enhanced properties. If the alloy is heated to a temperature at which the solute is completely dissolved, and then quenched, the rejection of solute atoms into the second phase that would result on slow cooling is prevented. In this metastable state, an attempt to return to equilibrium is thermodynamically desirable, since it removes strain in the lattice and leads to a lower free energy in the system. Such a process is called *ageing*. At room temperature specimens of many alloy systems of this kind produce atomic movements that cannot be seen, even under the electron microscope, but can be detected on X-ray diffraction photographs. At higher temperatures, further changes take place, more rapidly as the temperature increases. The subsequent precipitation that results can be stopped at various stages by cooling again, and a range of properties between those of the quenched and equilibrium structures can be produced. The duralumin series of alloys, based on aluminium with the addition of copper, are in this category, but there are now many more. Some rely on this feature for high strength at high temperatures, e.g. various alloys produced for gas-turbine components, such as the *Nimonics*.† Metallurgical textbooks should be consulted for details of these, since the previous warning on the stability of non-equilibrium structures applies very forcibly to precipitation-hardened alloys.

8.9. *Complex alloys*

Ternary, quaternary, and more complex alloys require special consideration, and are outside the scope of this book. Equilibrium in these alloys is difficult to attain, even in the laboratory, yet many of them are in normal commercial use. Steel itself contains, superimposed on the basic iron–carbon system, silicon, manganese, sulphur, and phosphorus, which are inherent impurities in the process of manufacture.

A reasonable picture can often be built up from the individual binary systems, e.g. a ternary system ABC will involve combination of the three binary systems: AB, BC, and AC. A solid figure replaces the binary two-dimensional one, and the effect of the third element C on the properties of the binary AB system can be estimated. Quaternary and more complex alloys are normally simplified by assuming that if D dissolves in one other constituent, say B, then the solid solution of D in B can be interpreted as constituent B. Experimental measurement of properties can be used to confirm the validity of such assumptions.

† Registered trade mark.

9

Non-metallic Materials

9.1. *General*

In Chapter 3, the effect of the crystal lattice and the various means of binding atoms together was described, with some reference to the resultant properties of materials, amplified in later chapters. The user of engineering materials, however, is obviously also interested in combinations of materials, particularly if a more desirable range of properties accrues from this. The metallic alloys discussed in principle in Chapter 8 are examples of this approach applied to the metals. In the present chapter, the same consideration will be given to non-metallic materials, both natural and synthetic. Detailed properties will not be given, for these are under continual improvement, and appropriate sources of information should be consulted. (Typical examples are shown in Table XII.) The aim in this book is to outline the nature of materials and the mechanisms of their behaviour in scientific terms.

9.2. *Timber*

This is a natural material, already highly heterogeneous, but lending itself to modification of properties by various treatments and methods of utilization. The two broad groups of timber are softwoods, from conifer trees (pine, etc.), and hardwoods, from the dicotyledons (oak, ash, etc.). The growth of trees involves two different additions to the thickness in each year, the so-called *spring wood*, and the later *summer*

TABLE XII

Properties of some non-metallic engineering materials

Material	Density ($g\ ml^{-1}$)	Electrical resistivity (at 20°C unless otherwise stated) (Ω cm)	Thermal conductivity (cal cm $degC^{-1}\ cm^{-2}\ sec^{-1}$)	Coefficient of thermal expansion (linear) ($10^{-6}\ degC^{-1}$)	Modulus of elasticity (10^6 lb in^{-2})
Aluminium oxide	3·8		0·07 (20°C)	5·4	50·0
Magnesium oxide	3·6	10^5 (1,000°C)	0·003 (50°C)	10·8	12·0
Silica (fused)	2·21	10^{20}	0·0033	0·5	10·4
Graphite	2·3	10^{-3}	0·157	3·5	1·0
Silicon carbide	3·17	2·5 (1,000°C)	0·029	4·5	—
Zirconium oxide	5·6	—	0·007 (100°C)	11·1	—
Beryllium oxide	3·01	—	0·51 (251°C)	7·5	—
Tungsten carbide	15·63	2×10^{-2}	0·19	5·0	102·8
Titanium carbide	4·5	5×10^{-5}	0·07	7·2	48·0
Glass: Pyrex	2·23	10^{14}	0·003	3·0	5–10
Glass: Soda	2·4	5×10^{11}	0·0015	8·5	
Building bricks	2·2–2·3		0·0015	3–9	—
Firebricks	2·1	$1·5 \times 10^8$	0·0025	4·5	—
Silica bricks	1·75	$1·2 \times 10^8$	0·002	12·5	—
Concrete	2·2–2·4	—	0·0025	10–14	2·0
Phenol formaldehyde resins	1·3	10^{12}	0·0004	42·0	0·8
Urea formaldehyde resins	1·5	10^{12}	0·0007	27·0	1·5
Synthetic rubber	1·5	10^{15}	0·0003	150·0	5×10^{-4}–10^{-2}
Polyethylene	0·92	10^{13}	0·0008	180·0	0·04
Polystyrene	1·05	10^{18}	0·0002	63·0	0·4
Polyvinyl chloride	1·25	10^{12}		65·0	0·02
P.T.F.E.	2·2	10^{16}	0·0005	55·0	0·05
Perspex	1·2	10^{16}	0·0005	90·0	0·5
Nylon	1·15	10^{14}	0·0006	99·0	0·4
Wood across grain	0·2–1·3	—	0·0001–0·00035	35–60	—
Wood with grain		—		3–5	—

or *autumn wood.* The concentric rings shown by a cross-section of a tree are caused by the different colour or reflectivity of these growths, but some trees do not show them (e.g. mahogany). A distinction is also drawn between the darker central portion of the tree (heart wood) and the outer sap wood.

The structure of wood is made up of long narrow fibres (tracheids) parallel to the trunk or branch of the tree, with patches called *pits,* through which water or sap can diffuse from one fibre to the next. The pines have resin ducts running parallel to these fibres, containing the resin so characteristic of these trees. Other softwoods have resin ducts; some have short resin-containing cells (Californian redwood).

At right-angles to these fibres and ducts are thin ribbons arranged radially, almost invisible to the naked eye in softwoods, and called *medullary rays*. These are prismatic-shaped cells containing complex chemical substances of nutritive value to the tree, and consequently also enjoyed by animals and fungi attacking from the bark.

Hardwoods contain some wood fibres in the form of water-conducting tubes called *wood vessels*, known as *pores* when viewed in cross-section. In the sap wood these contain air also. The medullary rays are much thicker in hardwood.

The heart wood in a tree is dead. The sap wood is the growing area; hence the water content is highest in the sap wood. The vessels and fibres in heart wood are, however, 'gummed-up' with various chemical products, and consequently are less receptive to impregnation by synthetic hardening materials. The main chemical constituents of wood are cellulose and lignone. The cellulose remains after treatment with weak acids or steam in the manufacture of wood pulp, and can itself be fermented to alcohol or dissolved in nitric acid or acetic acid to produce the esters that are a starting point for synthetic fibres (cellulose nitrate and acetate respectively). There are small quantities of tannin, oils, and other complex substances present in wood, all of which affect the properties.

The wood cells can take up a maximum of about 30% by weight of water, and swell in doing so. The wood becomes softer, more flexible, and easily shaped when saturated with water. Similarly when heated (as in steam) the wood fibres are more readily bent to reasonable curves.

The density of wood substance is usually about 1·5; timber of lower density (balsa) has more cavities per unit volume. Heavy woods such as ebony sink in water (density greater than 1) even when completely dried. Since wood stored under cover but exposed to the air will adopt the vapour pressure of moisture in the atmosphere, the water content of timber stored in this way varies very much with climate. In subtropical but dry climates, about 5% water remains, but in a temperate climate (Great Britain) about 15%. The shrinkage when dried is anisotropic, depending on the nature of the timber and its previous treatment.

The mechanical properties are obviously anisotropic, by virtue of the fibrous grain-structure. In tension and compression, the greatest resistance to deformation is along the grain. In shear, the greatest resistance is perpendicular to the grain. Every amateur woodworker can recognize the difficulty of using chisel or plane (shear forces) across the grain of timber. In bending, the stress is greatest when applied perpendicular to the grain, since the normal stresses are then along the grain. Timber used for beams is therefore cut with the longitudinal axis of the beam parallel to the grain as far as possible. Any change in grain owing to knots or other defects constitutes a weakness in this case.

Wood is a very superior material with respect to strength/weight ratio, having a higher value than steel. In tension, however, the greater strength of steel requires so much smaller cross-sectional area that the advantage of wood is lost. In compression or bending, provided that bulk is no disadvantage (as in building), wood may be most suitable. When vibrated, the high modulus of elasticity of wood, compared to other similar organic materials, produces strong resonance, and hence wood can be used as a sounding-board in a musical instrument. The strength is greater by a factor of 2 in the direction of the grain; in instruments thin sheets of wood are therefore used with the grain parallel to the plane of the sheet.

Various methods have been adopted to produce a relatively uniform grain-structure and so overcome the anisotropy of wood. Plywood consists of thin sheets glued together, with alternate sheets having the grain direction at an angle (usually 90°) to the other. This produces a more isotropic behaviour when stressed. If the glue is a water-repellent resin, then the shrinkage and warping of the wood due to changes in humidity is largely avoided. Blockboard, in which bars of rectangular cross-section are glued with their grain parallel to the tension direction, and top and bottom sheets with the grain at right-angles, can be substituted for certain applications, being cheaper than plywood for the thicker sheets.

Impregnated wood, in which synthetic resins are used to fill the cavities left by drying out water from the wood vessels, can be expected to resist humidity effects and to be somewhat stronger than the original wood. Synthetic wood produced from sawdust or wood shavings, intimately mixed with synthetic resins and then cured, has properties roughly half-way between impregnated wood and plastics with a wood filler. The mechanical properties are also intermediate, eliminating the anisotropy of the wood by breaking it up, and introducing more plastic flow into the hardened resin. Such materials can be formed into special shapes where required.

9.3. *Stone, concrete, and asphalt*

The 'aggregate' materials in this category comprise a wide range of mixtures. In each case an agglomeration of natural materials is intimately mixed together with a natural or synthetic binder, which then hardens and produces a strong component with a durable surface.

The natural stones are themselves aggregates. Igneous rocks, cooling from a melt, often consist of numerous crystals embedded in a matrix from which they were rejected as a result of the change in solid solubility on cooling. Sedimentary rocks, laid down on the floor of comparatively shallow seas or lakes, contain finely divided crystals embedded in a matrix, often containing hydrated silicates, or, in the case of limestones, a very fine carbonate precipitate. The properties of such materials are governed by this constitution and structure. The compressive strength is high, the resistance to thermal shock poor. The latter property results from the wide variation in thermal expansion possible between the different constituents, which causes thermal strains. The igneous rocks weathered into the fine material that became the sedimentary deposits partly as a result of thermal shock, assisted by rain, ice, and wind. Stone, therefore, can be used only in compression and at low temperatures. The surface deterioration due to weathering continues if the stone is exposed, and industrial atmospheres can increase the rate of attack, particularly when sulphurous gases are present.

To manufacture an aggregate to any desired shape by a more economic process than dressing stone, and in greater bulk quantities, the stone may be broken up and mixed with a cement to bind it together again. By this means, greater control over the binder, variation in the proportions of aggregate to binder, and a wide choice of aggregate, can be exercised. The natural stones vary widely in properties, even from one quarry, as a result of their origins. Compression strengths of sandstones vary from 2 to 8 ton in^{-2}; of limestone 2·5 to 12 ton in^{-2}; of granite 4·5 to 20 ton in^{-2}; and of basalt 10 to 30 ton in^{-2}. The value of the synthetic aggregate lies in consistency of properties and flexibility of operation. The two main binders are Portland cement, a hydrated calcium aluminium silicate, and bituminous asphalt, a viscous fluid originating from petroleum or coal distillation.

Portland cement is a silicate with a structure similar to the general silicate one described in Chapter 3, whereby calcium and aluminium atoms have substituted in the normal tetrahedral-type lattice of silica SiO_2. When finely ground the material can take up water of crystallization; several hydration reactions, many of them reversible, can be described. The crystallization of these hydrated compounds can produce strong bonds by adsorption on the surface of an aggregate; hence the cementing action is similar to that in clay, but much stronger, by virtue of the compounds present having a limited number of molecules of water associated with them. Heating drives off the water again, and the cement disintegrates. The proportion of water added in excess of that required for hydration performs no useful purpose except to lower the viscosity and cause the cement to flow better. The excess water content has to be dried out, however, and may leave pores behind if this is too rapid. Addition of water is consequently kept under strict control.

Bituminous asphalt is a complex organic material, probably composed mainly of long-chain hydrocarbons with some giant molecules bearing oxygen atoms at certain points. These viscous liquids can adsorb on to the surface of the aggregate and so bind it together. The change in orientation of the giant molecules prevents them moving relative to each other as they did before adding the aggregate, and so viscosity is enormously increased. Further additions of filler material complete this process, and a material behaving much more like a solid finally results.

The manufacture of bricks resembles the processes described above, but the bonding is usually vitreous in nature, i.e. amorphous-type structures are produced, as described for silica and the glasses in Chapter 3. The covalent bonding in glasses produces high compressive strength, of the order of 40 ton in^{-2}, rising to nearly 100 ton in^{-2} for pure fused silica. Bricks are often made from aggregates of old bricks ground up, cemented together with silicates. The latter may be either hydrated (clays) or of relatively low melting point (i.e. containing some impurities).

The essential properties of the aggregate depend on the close-packing of the load-bearing material, and the complete infilling of cement to bind it together. In Chapter 3, the close-packing of spheres was discussed, and the figure of 74% quoted for the volume fraction occupied by spheres of one diameter only (a, Figure 9.1). To occupy more of the space, spheres b of smaller diameter are required to pack into the interstices. This still leaves spaces and even smaller spheres c are necessary to fill these. Any remaining space may be filled by the binding material, to give 100= packing. Variations on this theme are possible. The 26% space left by spheres of one diameter can be filled by a mixture of binder and very small spheres, rather than using two intermediate grades of aggregate. This is often done in concrete mixing. The aggregate is rarely spherical in shape, and angular particles cannot fit together as closely as spheres, so that a mixture of gravel, sand, and cement is quite acceptable for normal concrete. If the best possible results for compression strength are required, and particularly for the highest density, two grades of gravel, plus sand and cement, are utilized. The proportions may vary from

1 cement, 3 sand, 6 stone (stone greater than 1 in. but less than 3 in.) for moderate strength in bulk, to 1 cement, 1 sand, 2 stone (stone less than 1 in.) for higher density and reinforcing work. In the latter case, a network of iron bars is arranged to take up tension stresses and prevent shattering of the concrete. In prestressed concrete, tension is applied to these bars while the mix is setting. On release of the restraint, the concrete is put into compression by the relaxation of these bars, and hence any tension applied in service must take up this compressive stress first, before a tension stress operates on the concrete. Concrete is attacked by sea-water and by certain salts that may be present in the soil, e.g. magnesium sulphate. Some oils attack

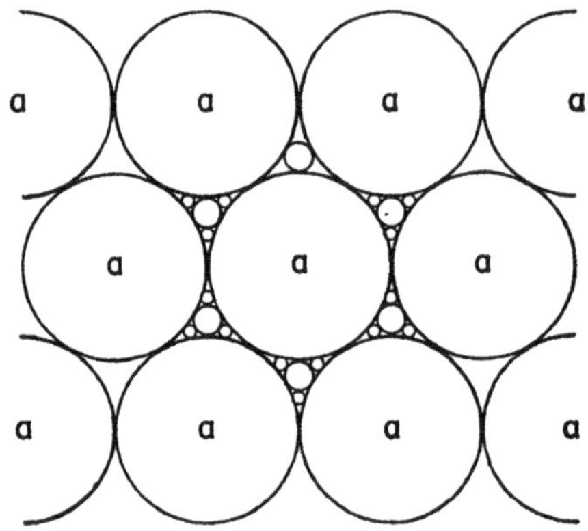

Ob oc

FIG. 9.1. Close-packing by three sizes of spheres, a, b, c (2-dimensional).

the cement, by combining preferentially with the calcium present. Concrete subjected to such environments requires to be very dense, and therefore impervious.

Similar arguments apply to asphalt-bound aggregates. The coarsest material may be gravel or stone chips, the intermediate size sand, and the very fine material dust-like particles of limestone, slate, or quartz. Road materials utilize this type of combination, and will withstand very heavy loads. The aggregate is brittle, of course, compared with the original binder, but can be softened by heat. It is therefore frequently applied hot.

Refractory bricks are expensive, and great care is necessary to ensure high density for application in furnaces holding molten metal. The grading of aggregate sizes is carefully controlled, and impurities are eliminated as far as possible to prevent softening at too low a temperature in service. The essential properties of compressive strength and high density apply as for other aggregates.

9.4. *Plastics*

Although, strictly, reserved for materials showing particular behaviour under deformation, the word *plastic* has become synonymous with any organic material manufactured by synthesis and shaped by plastic deformation. In Chapter 3, the covalent bond in organic chain and ring compounds was discussed, and the phenomena of polymerization. Van der Waals' bonds can exist between the individual molecules, so imparting a more rigid structure. The longer the chain or the more the branching, the stronger will be the Van der Waals' bonds (due to polarization). However, they are still weak compared with the covalent bonds of the individual molecules, and consequently the melting points of these substances are relatively low. Substances that undergo polymerization can form giant molecules, with strong covalent bonds replacing a number of Van der Waals' bonds that would be present if the individual units of the polymer were not condensed together. The size of the giant polymer molecule will obviously affect the melting point and therefore the mechanical strength, as with crystals. A change in molecular weight from 7,500 to 15,000 may increase the tensile strength from around 4,000 lb in^{-2} to nearly 10,000 lb in^{-2}. During polymerization, not all molecules will form to the same size, and consequently the properties will represent a mean value of the size distribution. The unit of the polymer is the 'mer' (Figure 3.37, page 87); in practice the variation in size of molecules may be from 1,000 to 10,000 mers per molecule. Molecular weights may be upwards of 500,000.

Addition polymerization does not occur merely by monomers sticking together automatically. The reaction, induced by heat, or pressure, or both, may require the assistance of a catalyst to first break open double bonds to obtain the unit 'mer'. Once this has been done, and considerable energy is involved, the units may add together until limited by the process of diffusion. Once a long chain has formed further additions can be made only at the ends, and hence diffusion is paramount. The ends must also be made stable, and this need happen only at one end initially. Figure 9.2 shows the breaking of a double bond to form the 'mer', followed by stabilization and subsequent growth. Hydrogen peroxide H_2O_2 and halogen atoms are commonly used for this, but other compounds, including organic ones, are possible.

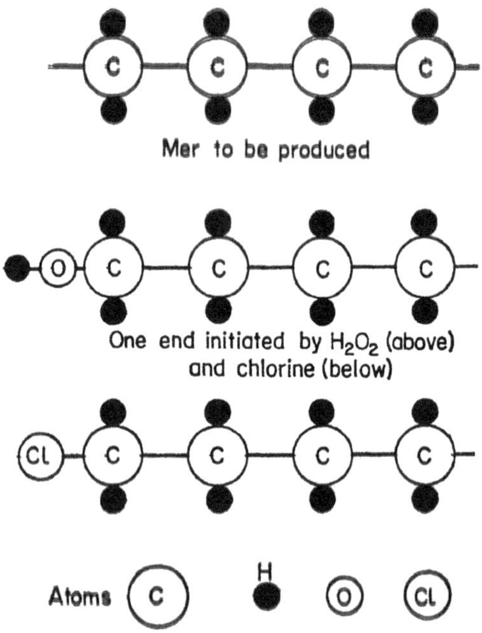

FIG. 9.2. Initiation of mers for polymerization.

In the presence of more than one type of monomer, a *copolymer* can be formed, in which individual units are formed from either monomer in a single molecule, resembling an alloy solid solution. The properties are then intermediate between the two extremes, and often vary in a manner not dissimilar to the metallic solid solutions mentioned in Chapter 8. Chemical reaction may take place between two different molecules, with the resultant joining together of molecules of one of the constituents by the other. In these cases, water (or a compound incorporating –OH) is often eliminated as part of the process, commonly known as *condensation* because of the bringing together by a reaction rather than addition. The properties of these polymers are vastly different from those of either of the separate constituents, since an entirely new compound is formed, in addition to a giant molecule. Many of the earliest plastics were of this type, e.g. *Bakelite*† formed from phenol and formaldehyde (Figure 9.3).

When polymer molecules extend in one main direction (linear or chain molecules) the cohesion is limited by the weak Van der Waals' bonds between molecules, and the entangling of the giant molecules. As the temperature rises, the entangling becomes more easily broken,

† *Bakelite* is a trade name.

and so do the Van der Waals' bonds; the material therefore becomes less viscous. In addition, depolymerization can occur, and the molecules may be broken up again into smaller molecules, although this happens only at high temperatures. These materials therefore become more plastic with rise in temperature, and are consequently described as *thermoplastic*. The addition types of polymer, in which giant molecules build up from a unit 'mer', are thermoplastic. The degree of branching of the molecule, and the resistance to depolymerization, determine the temperature at which they become less viscous and the rate at which plasticity increases with temperature.

When the polymer is formed by condensation reactions, a more

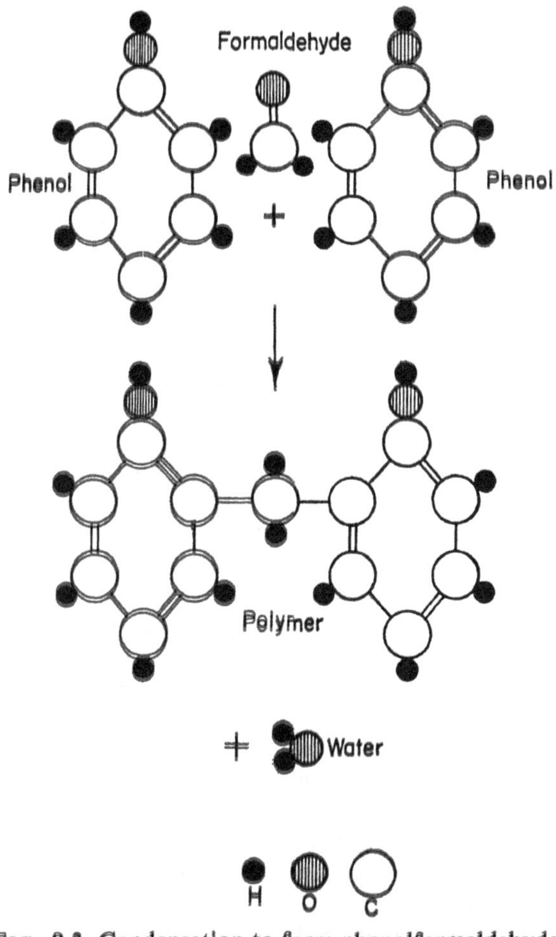

FIG. 9.3. Condensation to form phenolformaldehyde.

rigid framework is built up, which cannot be so easily broken. The whole structure can become in effect one giant molecule, having a three-dimensional framework of covalent bonds, and not requiring the introduction of Van der Waals' forces to hold the units together. The effect of temperature is then limited to thermal agitation of covalent bonds, and little plasticity results from this. Subsequently, continued rise in temperature produces oxidation or charring, and the structure disintegrates as a result. Such materials are designated *thermosetting*, because they assume a shape during manufacture but cannot be deformed readily merely by rise in temperature. These are the strongest plastics, but they are also relatively brittle compared with thermoplastics. Some thermosetting plastics have high electrical resistivity, a fact that has been reflected in their use in electrical fittings.

There are several possible means available for modifying the properties of thermoplastic materials. Mention has already been made of the branching of long chains, giving increased Van der Waals' forces and interfering with slip during deformation. The simple straight-chain hydrocarbon, such as polyethylene (Figure 3.37, page 87), cannot be so resistant to deformation as a similar chain compound with extra atoms incorporated (polyvinyl chloride) or the replacement of some hydrogen atoms by larger ones (polystyrene). These effects are illustrated in Figure 9.4. Molecular orientation as a result of polarization forces (affecting the Van der Waals' bonds) takes place when the material is being fabricated; as a result, the dipoles, where they exist, tend to come together. The intertwined molecules then have portions where the chains run parallel to each other, with strong electrical forces (Van der Waals'). When deformed in a particular way, these parallel portions of molecules align themselves in the direction of mechanical strain, and so produce orientation effects of great importance in the fibre industry. The parallel molecules are called *micelles*, and are characterized by nylon and the cellulose (acetate) compounds used in synthetic fibres. In die-casting and extrusion, which are commonly used for fabricating thermoplastics, the orientation tends to be along the lines of flow. The rate of cooling then determines whether 'recrystallization' can take place, modifying the intertwining of molecules or the arrangement of strong dipoles in parallel. Where dipole moments are strong, then the orientation along flow lines is not detrimental, and rapid cooling is possible. Where there are no dipoles, or only weak ones, then slow cooling allows reorientation and intertwining of the molecules to form micelles.

For more complex polymers the isomerism referred to in Chapter 3

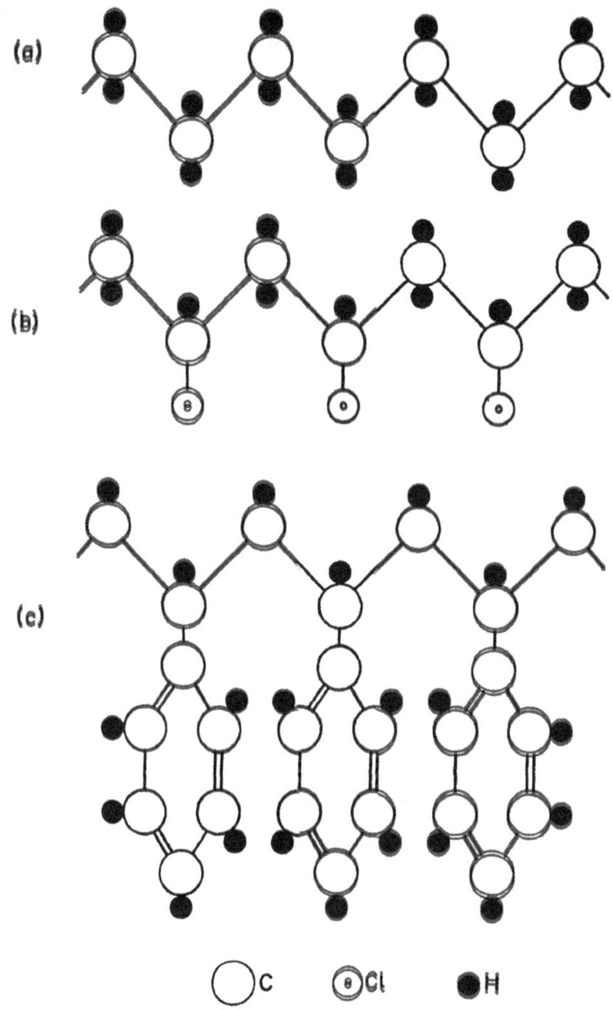

FIG. 9.4. (*a*) Polyethylene. (*b*) Polyvinyl chloride. (*c*) Polystyrene.

(page 87) becomes more important in determining properties. A more complex group in a chain molecule may be aligned on one side or the other relative to the hydrogen atoms, to double bonds in carbon, or to other atoms present. Rotation of the complex group may not be possible if it is associated with double carbon bonds, and bond angles may be modified only slightly. It is therefore possible to have two polymer molecules of identical composition but different structural orientation, with obvious differences in properties. The classical case is that of natural rubber, which chemically is quite a simple structure, a polymer of isoprene, whose formula may be written

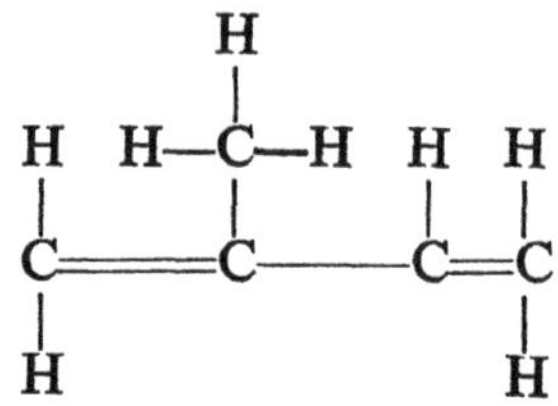

When isoprene is polymerized, the double bonds at each end are broken, to form the 'mer'. A double bond then exists between the central two carbon atoms. The mer of isoprene is then

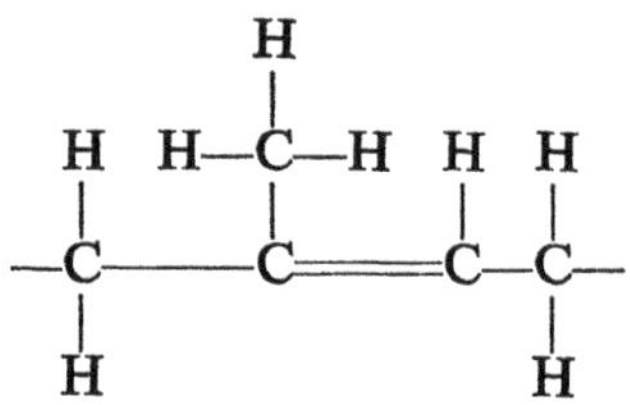

and polymerization takes place by addition at each end. The ends of the mer are symmetrical, but the central portion can have the two groups, a methyl (CH_3) group, and a hydrogen atom, on the same side (*cis*) or on opposite sides (*trans*) of the chain. When they are on the same side (*cis*) then the material is natural rubber, polyisoprene. When on the opposite sides (*trans*) the mer is

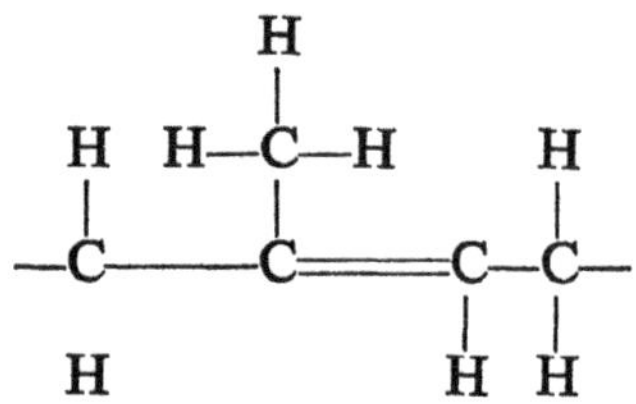

and the polymerized form is gutta percha. Natural rubber is an *elastomer*, i.e. a polymer that recovers the change in shape caused by a deforming stress, once the stress is removed. Gutta percha, however, is much harder and stronger, but not elastic. The main reason for the difference is the possibility, in natural rubber, of the molecule's coiling up on itself, in an endeavour to occupy the minimum surface area. This is feasible because in the *cis* form of the polymer both groups about the double bond are on the same side, leaving room on the other side for bending of the single bonds in the chain without bringing two hydrogen atoms too close together. (The hydrogen atoms here at the ends of the mer.) In the *trans* version, the bending is not so easy in either direction, and so does not occur. When stretched, a coiled molecule must first be pulled into a long chain before the bonds are themselves stretched. This accounts for the high elasticity.

The mechanical properties may be further modified by the phenomena of *cross-linking*. Here, connections are made somewhere along the molecular chain, at points where unsaturated (double) bonds occur between carbon atoms. Obviously it is impossible to add on any bridging atom or molecule unless there are unsaturated bonds to satisfy. One such possibility occurs in rubber, since there is a double bond at every fourth carbon atom in the polymerized molecule. Oxidation takes place, even at room temperature, in natural rubber, and the oxygen atoms linking across between two of the former double bond carbon atoms in two separate molecules:

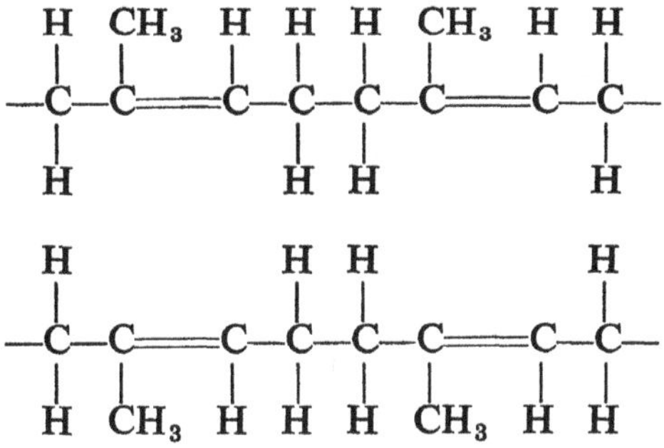

becomes

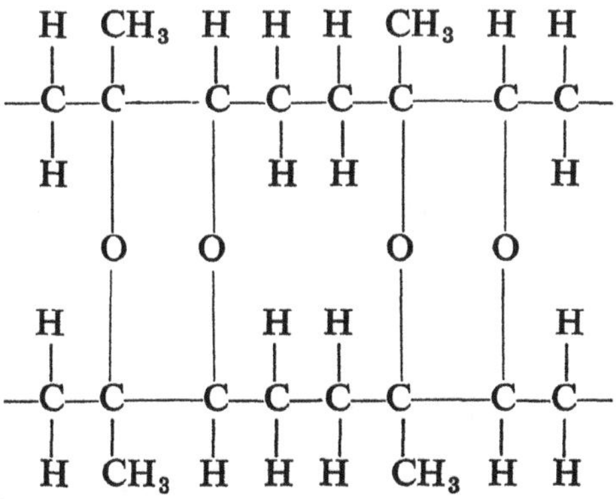

The rubber is then unable to deform elastically as it did previously and becomes hard and brittle (*perished* is the colloquial term).

Cross-linking can be advantageous, as in the vulcanizing of rubber. The high elastic modulus of rubber is accompanied by some slip and plastic deformation unless there is some cross-linking, and only a small bridge is necessary to do this satisfactorily. The original invention of GOODYEAR incorporated sulphur for this purpose as an addition to natural rubber. Heating completes the process by breaking the double bonds and allowing sulphur to form bridges exactly similar to those shown for oxygen above. About 2–5% sulphur fixes enough molecules to give good elastic properties, but any excess over this produces so many cross-linking points that the material loses its elasticity, and becomes *ebonite*, a hard resistant solid. This process is known as *vulcanizing* when applied to rubber. A similar well-known example of cross-linking is the drying of paints based on linseed oil. Oxygen atoms here provide the cross-link between individual polymer molecules, and the paint film becomes hard and less permeable. This process takes time, for oxygen diffusion through the first hard film on the surface becomes increasingly difficult. Thin films are therefore more readily hardened than thick ones.

The effect of branching in long-chain molecules has been referred to in connection with increasing the mechanical strength. Apart from choosing molecules that have branched chains, the possibility now exists of artificially condensing molecules to produce quite distinct branched molecules. The mechanism is similar to that for condensation polymerization and involves the elimination of water or similar by-products. In effect, a side atom (often hydrogen) from one chain and an end group from another (often –OH) are removed, and a carbon–carbon bond introduced in their place. Heat, pressure, or catalysis is necessary to perform the reaction, which may become extremely important in the development of plastics.

Very few plastics are utilized in the raw state. Most have additions made to modify the properties, to suit the applications. Filler materials, usually cheap by-products from other industries, are used to vary the density, opacity, colour, heat resistance, or mechanical strength. Clay, chalk, sawdust, asbestos, paper, peat, and zinc oxide fall into this category. Fibres may be added as reinforcement, either as laminates or by intimate mixing in powder form. Pigments or dyes are required for colouring, and plasticizer materials of all kinds are added to improve plastic behaviour. Economic considerations may determine the range of applications of plastics, and the value or disadvantage of modifications to properties. The normally high resistivity of many thermosetting plastics makes them eminently suitable as electrical insulators. The covalent bond gives rise to the low electrical conductivity, but the polarizable nature of the molecules also offers dielectric properties. Some organic salts are formed between metals of groups I or II of the periodic system (Table II) and polymerized molecules bearing acid radical groups at the ends of the chain. Many of these have been developed for *ion exchange*, i.e. in water-softening or for separating metals from a mixed aqueous solution. Here, the metal ion of the resin is exchanged with the metal ion in the solution, and subsequently restored by backwashing with a salt solution to reverse the procedure. Some resins are anion exchangers, and some cation, i.e. they can pick up metal ions or acid radicals. The basic structure of the polymerized material is not affected, only the exchanged atoms or groups.

9.5. *Ceramics and glasses*

In Chapter 3 the structure of a number of compounds was discussed, ranging from the ionic compounds to the completely covalent ones, and including unusual structures such as layer lattices. The individual properties were referred to in terms of the type of structure. The original concept of a range of mechanical strength in minerals was put forward by MOH, who drew up an order of merit on a scale of hardness. On this scale, each mineral could scratch the one below it, and be scratched by the one above. Talc was lowest in the scale, and diamond the highest. When arranged in order of merit by structure, the scale reads as follows:

Layer lattice (e.g. talc, graphite)
Ionic lattice (e.g. sodium chloride or calcium fluoride)
Silica tetrahedra (e.g. quartz)
Complex ionic–covalent lattice (e.g. silicates)
Covalent lattice (e.g. diamond, alumina)

There are variations on this order, produced by mixtures of ions and by structural defects, but apart from overlapping of near neighbours the scale is approximately correct. The layer lattices are easily sheared, since they possess directional properties, weak Van der Waals' forces operating between the layers. They are therefore used as lubricants (talc, graphite) or as plasticizers (clay). In the remaining types of lattice, slip is not easy, as explained previously, and plastic deformation is consequently very restricted. These materials fail mostly by normal stresses, cleavage occurring along crystal planes, with a brittle, i.e. a non-ductile, fracture. Since, in compression, the cleavage cannot take place in the same way, these highly bonded lattices exhibit high strength; their main use is consequently as compression members of components. The melting points of many of these compounds are high, and the oxides in particular are not normally affected by oxidation, so that they retain mechanical strength in compression to quite high temperatures. This accounts for their use in refractory bricks and linings. Clay is carefully dehydrated in firing, and loses its normal layer lattice behaviour, since this is due largely to water molecules between the layers. The fired material consists of ionic bonds in some planes and covalent bonding replacing the weak Van der Waals' forces between the original layers, this bond change giving rise to increased shear strength, less affected by temperature. Oxides, carbides, and borides can also be used as cutting tools or grinding wheels, retaining hardness even at high temperatures.

Amorphous materials, such as glasses, are already in a state of strain, since they have been supercooled and not allowed to crystallize. Application of stress breaks some of the atomic bonds in these lattices, those with the highest residual strain in the first instance. Other bonds then take up the stress, and are broken gradually when the stress across them reaches the appropriate level. This gradual breaking of bonds is characteristic of *viscous* flow, where flow is proportional to time. The ratio shear stress/rate of strain is equal to a constant, the viscosity. At room temperature, the rate of strain is low, but at high temperature it increases markedly. Such substances therefore do not have a true melting point, but exhibit a softening range in which viscosity gradually decreases. These amorphous structures are therefore strong and brittle (non-ductile) but can deform as temperature rises.

The electrical properties of these materials were discussed in Chapter 6. The ionic lattices have a measurable conductivity, but the rate of diffusion of ions largely determines the quantitative value. Since diffusion rate increases at higher temperatures, so does the conductivity of an ionic lattice. The polarization of ionic lattices, giving rise to ferroelectric properties, may be important in electrical applications, such as radio, television, and computers. Here the electrical properties are paramount and no structural strength is required.

The main disadvantage of the ceramics and glasses is their poor resistance to thermal shock. This was discussed in Chapter 5, when the various parameters used to assess this property were referred to. Since these substances are much used in high-temperature applications, the nature of thermal shock largely determines the necessary precautions to be taken in using them, or to be considered in design calculations incorporating them.

The thermal shock arises from the strain introduced by differential thermal expansion caused by the setting-up of a temperature gradient.

Such strains may be high enough to cause cracking, as with a thick-walled glass vessel suddenly filled with boiling water. In a thick wall the cracks may not penetrate to the surfaces, but they nevertheless represent discontinuities. The thermal conductivity across the surfaces of these discontinuities is obviously reduced, and the temperature gradient accordingly rises. This produces further strain by differential expansion, which may propagate the original crack, or cause the production of fresh cracks. In either case the original effect is exaggerated, and if again the cycle of reduced conductivity, increased temperature gradient, and increased strain repeats itself there may finally be distinct separation into two units, with a macroscopic gap between them, and an even greater temperature drop. This situation can arise in refractory bricks, and leads to *spalling* of the hot surface, which may separate completely and drop into the furnace hearth or the pool of molten metal or slag. The process may then be repeated on the freshly exposed surface.

One solution to this problem, used in furnaces, is to insulate the rear (cooler) face of the brick, to prevent loss of heat and so allow an overall rise in temperature throughout the facing brick. In the milder temperature gradient which results in the facing (hot) brick, the spalling strains may be much reduced, and cracking become negligible. Inevitably, however, the hot face of the brick must reach a higher temperature than before, as the heat losses have been reduced. Provided that this hotter working face is not too near the softening temperature, this is not harmful. In the open-hearth steel furnace having a roof of silica brick, the roof is already at a very high temperature (1,700°C), and may actually drip molten silica during operation. Spalling (by the mechanism described above) would, however, remove so much greater quantities of material than slow melting that the whole roof would eventually collapse. (Modern furnace design incorporates mechanical aids to assist the bricks, by transferring some strain to a suspended framework of metal.) As a furnace material, silica is also complicated by the existence of three allotropic forms (Chapter 3) and the variation in thermal properties between them. Thermal strains from this cause are set up in the firing of silica bricks and later in service, quite apart from the service strains referred to above.

9.6. *Cermets*

The poor resistance to thermal shock and brittleness of ceramic materials, described in the previous section, should be improved by the incorporation of a good conductor or a material showing greater ductility. The use of a metal–ceramic mixture for this purpose led to the name 'cermet' where the ceramic portion was not limited to oxides, but included carbides, silicides, borides, and, more recently, nitrides. Carbide tool bits, cemented by a ductile metal, come within this category, and, strictly, reinforced concrete should be so regarded. The high melting point of the ceramic portion of the structure helps to retain high strength at high temperatures, at which the metal alone would be far too plastic. Obviously, the behaviour of the cermet will depend on the proportions of metal and ceramic, the particle size and shape, and the general distribution of the minor constituent in the matrix. Numerous permutations are therefore possible, from a metal matrix with ceramic particles embedded in it, to a ceramic body in which metal fibres, platelets, or spheres, are distributed more or less uniformly. Apart from this concentration and dispersion variable, the nature of the interface between metal and ceramic portions is of considerable importance, both for strength and the optimization of other physical properties.

The concentration factor can be considered by simple mixture rules. The contribution to properties of each constituent is presumed to be additive, and in proportion to the volume occupied by each. Such a simple rule can apply only to simple properties such as density, and where there is reasonably uniform distribution of the two constituents. A layer structure, for example, would obviously have variable properties in the direction of the layers as against those normal to the layers. Spherical particles embedded in a matrix will behave differently from disc-shaped or rod-shaped particles. If spherical particles have properties superior to those of the matrix, they contribute little to the overall property, since the matrix is coherent and plays the major role. In the reverse case, the square root of the parameter for the matrix will be involved. For other shapes, the indices involved are different.

MAXWELL derived a more exact formula, incorporating the two main possibilities, whereby either the matrix or the dispersion plays the major role. The logarithmic mixture rule involving the geometrical mean of the properties is perhaps the most practical to use. This is most simply applied by plotting the parameter concerned on a log scale and the volume fraction of the minor constituent on a linear scale. This rule applies only where there is a matrix dominating the property, as distinct from a simple uniform mixture, for which the simple rule applies (as in density). Figure 9.5 illustrates the two rules for typical materials.

Although the mixture rules are of value in predicting likely properties, the nature of the interface between the constituents plays an extremely important role in cermets, where the bonding between the two different constituents is so markedly different. Some materials are controlled by mechanical bonding, relying on the interlocking shape of particles. Others depend on chemical reactions at the

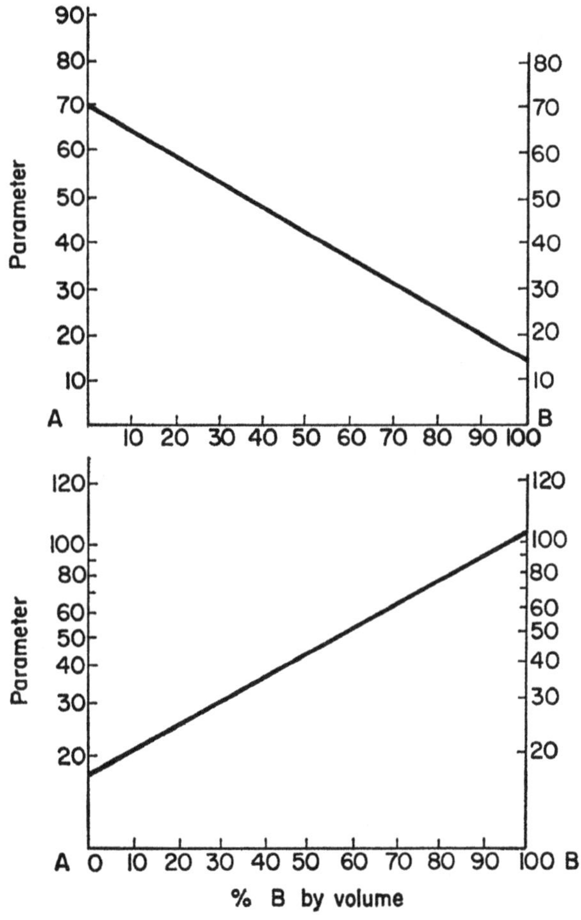

FIG. 9.5. Mixture rules. (*a*) Simple. (*b*) Logarithmic.

interface, whereby the products may or may not be soluble in either or both components. Van der Waal's bonding will obviously be weaker than ionic or covalent chemical bonding, but the interface may be so thin that assessment of the bond strength is based on a parameter incorporating only the surface layers, i.e. the interfacial

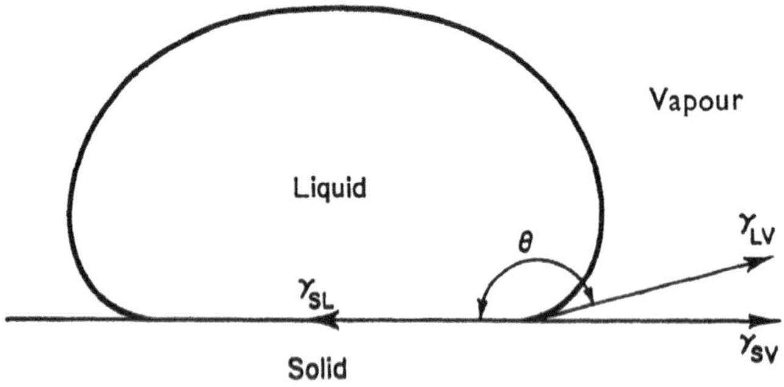

FIG. 9.6. Surface forces acting on a liquid drop resting on a solid.

energy γ. Figure 9.6 illustrates the model used, that of a liquid drop resting on a flat surface of solid. The relation between the surface energies is given by the equation

$$\gamma_{SV} - \gamma_{SL} = \gamma_{LV} \cos \theta$$

When $\theta = 0°$, complete wetting results, but here γ_{LV} must not be greater than γ_{SV}. In cermets, this latter requirement rarely holds. Values of θ up to 90° can be considered to represent satisfactory wetting for the purpose of bonding, and variations in γ_{SL} particularly are concerned in metal–ceramic bonding. The presence of oxide films interferes by increasing γ_{SL}; chemical reaction to dissolve the oxide film, or at least reduce it, hence serves a useful purpose in aiding wetting. The formation of oxide films during sintering of metal–ceramic mixtures is avoided by conducting the operation in a vacuum or in an inert atmosphere, or by introducing a reducing gas such as hydrogen.

Recent work has indicated that tensile strengths above 200 ton in^{-2} can result even with poor wetting. When wetting is good, strengths can rise to over 1,000 ton in^{-2}. The dispersion of the constituents can display a far greater influence on the nucleation and propagation of cracks under stress than on the basic strength, and a thin film of metal surrounding uniformly dispersed ceramic grains is generally recommended. The introduction of small quantities of a second oxide, which will dissolve in both metal and ceramic to a small degree, has been used to assist wetting and bonding. The chromium–alumina cermets contain little chromic oxide, which produces a solid solution with the aluminium oxide and dissolves slightly in the metal. It is believed that reduction in γ_{SL} occurs. Titanium is a popular metallic addition, as is zirconium. These metals tend to reduce the basic oxide of the mixture, the resultant TiO_2 or ZrO_2 then dissolving in the ceramic. The excess titanium or zirconium dissolves in the metal component, and good wetting and bonding result. Obviously this is limited by the solubility restriction to particular systems, but both titanium and zirconium have proved to be widely applicable in this connection. Intermetallic compounds have been used in a similar fashion; so have hydrides that decompose to produce oxygen-free bonding surfaces.

Differential thermal expansion between metal and oxide cannot be too great if thermal stresses are to be avoided. Differences of the order of $3–5 \times 10^{-6}$ $degC^{-1}$ have been quoted as acceptable, and 10^{-5} $degC^{-1}$ as totally unacceptable. Much depends on the dispersion in this consideration. Fine particle size and uniform distribution is favoured, since any thermal stresses due to differential expansion are then spread over the whole structure.

The properties of cermets so far developed have been, on the whole, disappointing. The problem of oxidation at high temperatures, which has slowed the development of the high melting point metals and their alloys (tungsten, molybdenum, tantalum, niobium) is also a factor with the cermets. Although the oxide is resistant, the metal may be oxidized, particularly since the high melting point transition metals are commonly involved. Carbides and other compounds may also be susceptible to high-temperature oxidation, so that other additions may be necessary, either to the cermet body as a whole or to the surface layers as a protective coating. The resistance to thermal shock remains a major disadvantage. The anticipated improvement in thermal conductivity by the addition of a metal to an oxide is less effective than the mixture rules suggest, and consequently thermal strains are high, though generally less than those for the pure oxide on which the cermet is based. Carbides often show better thermal conductivity in the pure state, and may be improved further by metal additions. Nevertheless, thermal shock remains a problem, judged by any of the parameters referred to in Chapter 5, or by empirical tests. Even the criterion based solely on mechanical strength shows that the problems of nucleation and propagation of cracks can seriously affect the resistance to thermal shock. The cermets designed solely for mechanical strength have not been entirely successful, mainly because the brittleness of the ceramic portion is largely carried over into the cermet mixture. Here again, the mixture rules may present a totally false picture, or indicate substantial improvement that is not confirmed by experiment. It seems likely that the bonding of the compound to the metal may be responsible for this variation, and bonding is by no means fully understood.

Recent improvements in oxide-base cermets depend on the addition of a second oxide. This may be partially soluble in the base oxide and atoms of the metal portion of the cermet may substitute in some sites on its lattice. Carbides continue to show improved properties when bonded with a wide range of metals, again perhaps involving better bonding by partial solid solubility or the formation of a mixed carbide at the interface. Borides tend to have low ductility, as do silicides. The latter, with nitrides, have only recently begun to be thoroughly investigated.

One considerable success of cermets has been in friction and wear-resistant applications, for brakes, bearings, and seals. Cermets can show an increase by a factor of 2 in friction values over sintered metals, and almost 50% increase over asbestos-based materials. Both oxides and carbides have been used for this purpose, bonded with nickel, chromium, or molybdenum, singly or in pairs.

Nuclear reactor applications of cermet materials have been based on fuel requirements, as in the dispersions of austenitic steel with uranium dioxide, or the use of oxides or borides in steel for control rods. The carbides appear to offer most promise, particularly when uranium–plutonium based fuels are considered. The fissile material oxides have low conductivity, and hence develop exaggerated temperature gradients in fuel elements. The temperature at the centre of the element may rise to the melting point, whereupon conductivity improves, the material solidifies, and in doing so cracks and creates porosities behind. The conductivity drops owing to the cracks and the pores, and melting occurs again. This cycle of events is repeated, and the material breaks up inside the fuel can. The addition of metal does not improve the conductivity sufficiently, nor the thermal shock resistance, and the economic disadvantages of the reactor system may increase.

The development of ceramic and cermet materials may be in its early stages, but the disadvantages discussed above must be minimized before full use can be made of their high melting point, high strength, and corrosion resistance.

10

Service Factors

10.1. *General*

The effect of atomic structure, the crystal lattice, and the possible combination or mixture of materials on the resultant properties, has been discussed in the previous chapters largely in terms of pure substances prepared in an ideal manner. Engineering materials do not or cannot always employ the purest of substances, and the methods of manufacture may be far from ideal, or largely restricted by economic considerations. Where appropriate the effect of impurities has been referred to; in this final chapter some of the service factors that modify the properties as a secondary influence will be discussed. Not only the fabrication of the material but the environment of storage before use, the service environment, and side issues specific to particular applications can upset the anticipated life and lead to wrong conclusions on the suitability of materials or the mechanism of their behaviour. These several factors will be dealt with in order of their likely appearance, i.e. from origins to manufacture, and then to service conditions.

10.2. *Heterogeneity*

Materials are rarely perfectly homogeneous, either in composition or properties. The presence of impurities may not always occur in a random distribution, and even the sampling techniques used for analysis may be affected by a statistical distribution of errors. In most cases, these heterogeneities of composition or apparent analysis are smaller than the effects of fabrication heterogeneity, but in semiconductor materials this is not necessarily so. Final assembly testing for quality may be necessary, since the net effect of several different variables can sometimes cancel out and not interfere with the designed usage.

In a number of materials, mixtures are used deliberately in order to obtain an optimum range of properties. The distribution of individual constituents is then important, as in the cermets. The distribution of physically separate particles may be even more vital, as in the aggregate mixes such as concrete and asphalt. Plastics are rarely prepared pure, but contain many added filler materials. Polymers may contain unpolymerized monomer molecules, or breakdown products. These chemical factors give rise to an average set of properties in the bulk material, which may be only slightly deleterious to performance, and can be taken into account in design studies. Equilibrium may not be reached in chemical reactions, not because this is impossible, but as a result of economic balance. The time required to attain equilibrium in certain reactions is so long that even a laboratory experiment may fail to reach completion. In other fields, equilibrium is deliberately avoided, in order to manipulate the metastable state, or to vary the transformation rate or product. Such examples are common in the metallic alloys for precipitation-hardening and in the hardening of steel. Subsequent secondary reactions to approach the restoration of equilibrium may introduce time as a variable, particularly if excessive rise in temperature is to be avoided.

Physical heterogeneity may be inherent, as in anisotropy, or result from variations in solidification rate or direction, or mechanical strain, during manufacture. Mechanical strain is very important, and is discussed separately below.

Metals are notoriously heterogeneous, both chemically and physically. Until comparatively recently, lack of understanding made the purification of some reactive metals difficult, but economic considerations largely govern the decision to adopt a purification process on a commercial scale. Steels, for example, are really most complex, since the normal method of manufacture through the blast furnace and steelmaking processes inevitably leaves a residue of phosphorus, sulphur, and silicon, apart from valuable alloying elements such as carbon and manganese, to say nothing of oxygen, nitrogen, and hydrogen. The effect of the solid elements can be reasonably accounted for, but insidious effects can arise from hydrogen and nitrogen embrittlement. Hydrogen can cause a large turbine rotor to disintegrate by forming hair-line cracks as a result of the build-up of hydrogen molecules formed by diffusion of atomic hydrogen into voids. Nitrogen may be in a metastable solid solution that later undergoes precipitation, sometimes with the appearance of heterogeneous ductility during fabrication—as in the pressing of sheet steel into motor-car bodies or the casings of domestic appliances. Methods of analysis, and of removal, of these deleterious gases are now much more reliable, but the dangers still exist. Many metallic properties other than strength or ductility are affected by small traces of impurities. Reference was made in Chapter 6 to the effect on electrical conductivity of impurities in copper. Only one atom of arsenic or antimony in one hundred of copper increases the resistivity by a factor of 5. Since copper is the cheapest of the really good conductors (silver, copper, gold), there is every incentive to refine the metal for electrical purposes, and electrolytic methods are frequently incorporated for this reason. All melting operations need to avoid over-oxidation, for copper oxide dissolves in the metal and affects both strength and conductivity. The oxide can be removed at the end of the operation by providing a reducing atmosphere (poling). The product is frequently designated 'O.F.H.C.', i.e. oxygen-free, high conductivity.

Metals and alloys are physically heterogeneous by virtue of their preparation, either by solidification of a liquid or the sintering of a pressed powder. The grain size of the cast material and the particle size of the sintered product exercise a remarkable influence on properties, particularly the mechanical ones. In Chapter 4, the grain-boundary lattice imperfections were described in terms of the dislocation movements. Similar conditions exist in the sintered metal, where diffusion takes place to decrease the size of the initial pores between particles, and recrystallization occurs across the boundary to varying degrees, according to time and temperature. Commercial metallic materials therefore behave as a collection of separate crystal grains, modified by the properties of the boundary, and would exhibit anisotropy but for the fact that the polycrystal contains many grains and a random distribution of orientations. Any departure from this, by grain growth and/or alignment of particular orientations in a more favourable direction, leads to increased variation of properties in different directions. Segregation may occur in solidification, by virtue of varying melting points of constituents present, causing further heterogeneity. In fabrication, rotation of crystal grains can take place, and, in the absence of annealing to recrystallize the material, anisotropy of texture results. Solidification itself can lead to anisotropic arrangements of crystal grains, especially when severe temperature gradients are set up, as in a chill mould. Porosity, due to grain boundary mismatching of different orientations in adjacent grains, or to segregation effects, can not only reduce the level of important properties, but act as a stress concentration, or a 'sink' for gaseous impurities such as hydrogen. In fabrication, mechanical defects can be set up by oxide scale on the surface being forced into the skin of metal, or by non-uniform deformation. In rolling, the rolls drag the skin of material through the gap faster than the central portion and so build up tensile stresses through the thickness. Wire-drawing holds back the skin compared with the centre, and so gives the reverse stress distribution. Extrusion may give rise to defects owing to non-uniform flow, and may cause oxide skin to be forced through the die with the metal. Such defects are well understood, but the human factor in such operations still plays a large part in their elimination.

Defects and variations in properties arise in many other materials for very similar reasons, and by virtue of mechanisms similar to those described for metals. Some materials absorb water from the atmosphere, by virtue of porosity in the body of material, or the special properties of the surface. Even if the absorption is uniform, then heterogeneity still results, since the properties are now those of a material containing an additional impurity. Heterogeneity plays an obviously important role in the *actual* as distinct from the *inherent* properties of materials. Quality control in commercial practice endeavours to minimize any disadvantages and to utilize any advantages, within the obvious restrictions of economics.

10.3. *Residual stresses*

When a material is prepared by solidification from the liquid, as with cast metals, glass, and some plastics, the solid forming at the mould walls, or whatever heat-transfer surface is made available, begins to contract as its temperature falls, and hence tries to shrink on to the still-liquid interior. The pressure of the liquid prevents this, and the solid shell is sufficiently plastic at the high temperature involved to deform under the pressure. Later, the centre begins to solidify within a solid shell that has remained expanded by the pressure, and hence could contain a larger volume than is now solidifying within it. The shrinkage of the centre now exerts a compressive force on the skin, but the skin is much cooler and much less plastic than it was, and so cannot deform so readily. When solidification is complete, therefore, the centre may be under high tensile stresses. If the body is reheated, the material may become plastic again, and, if time is allowed, deformation takes place with relaxation of the stresses. The subsequent cooling must be low, as otherwise the stresses can be restored again. This treatment is known as *stress-relief annealing*. Glass, and other brittle materials, may contain surface cracks as a result of stress relaxation during the original cooling, and these cracks propagate with time unless annealing is carried out soon after shaping of the liquid glass. Metals are often annealed before fabrication from ingots; in the case of steel this also equalizes the severe temperature gradient remaining after the mould is removed, as well as annealing the residual stresses. Some liquid might otherwise still exist in the centre and be squirted out in rolling or forging.

The stresses set up during cooling of an ingot are reasonably uniform in magnitude and distribution, since the shape is geometrically simple. In castings, whether of metal or any brittle material, there will exist changes in cross-sectional area, and junctions between heavy masses and smaller ones in a complex shape. Additional stresses result from the more rapid cooling of the lighter sections than the heavy, so that not only are surface stresses present in the skin, but a complex pattern of stresses may be set up over the whole volume of the cast shape. In joining processes such as welding, metal is cast into the prepared joint gap, and solidifies in the same way as an ingot, but more rapidly, owing to the severe chilling effect of a large mass of cold metal on a relatively small volume of liquid. In welding, the parent metal is in addition raised in temperature by the heat of welding, and hence expands. The structure may restrain the expansion, by virtue of its shape; hence, on cooling, tensile stresses are left in the heat-affected zone exactly as with the ingot skin. Residual stresses from welding can be very high, certainly almost equal to the yield point at room temperature. This is largely because the maximum stress is being exerted at the time when temperature is falling and elasticity decreasing, so that complete relaxation is not possible unless special precautions are taken.

Mention was made in the previous section of stresses set up in fabrication by rolling and drawing. The stresses here can be high, but are not normally as severe as those from casting or welding, unless very severe cold-working is carried out. The more plastic the material, particularly at room temperature, the more likely is it that residual stresses due to casting, welding, or fabrication will be relatively small. If they do exist, the principle of stress-relief annealing remains the only effective treatment to remove them. They can be minimized, of course, at the time of their production, by lowering temperature gradients or pressure gradients, and allowing time for relaxation during cooling or working. Temperature gradients are reduced in casting by employing moulds of poor heat transfer, and in welding by preheating the parent metal or post-heating immediately after welding. Pressure gradients are limited by reducing cross-sectional area in smaller steps.

The main problem arising from residual stresses is the decision on the real need for their removal in many cases, since stress relief costs money, and in some instances the structure may be so large that relief is well-nigh impossible. Since the maximum temperature of annealing must be high enough to lower the yield stress so that plastic flow can take place, there is a danger of affecting the properties permanently, particularly where a metastable structure may be involved. Steels are susceptible to microstructural changes at the temperatures required for stress relief, and for this reason relief is often avoided. The argument for and against the necessity for removing residual stress requires a knowledge of the stresses that will operate in service. In many cases there are constructional stresses, e.g. those set up by the suspension of pipes from the building in power stations and chemical plant, owing to the weight of the pipes, or the connection of one component to another. To these must be added the operational stresses of service, e.g. the steam pressure in power-station piping, or the rotation or reciprocation of moving machinery. If residual stresses are to be added, the resultant effective stress system may cause local deformation at some point. If this leads to relaxation of stress without further failure, all is well. If, however, a stress-concentrating crack is created, then even if relaxation does result, the operational and constructional stresses may together continue to propagate the crack and cause catastrophic failure. There is also a possibility that microstructural changes may take place under the severe stress system, even if cracks do not form. Subsequent conditions may again lead to rapid deformation in a particular area, with catastrophic results.

Residual stresses can also arise as a result of transformations in alloys on cooling. These are normally microscopic in distribution, but their summation can cause macroscopic effects, as in the quenching of steel for heat treatment. Martensite, the metastable structure produced by quenching austenite, is less dense than the parent structure from which it is formed, and so tries to expand during transformation. This is prevented by the untransformed centre, since the cooling proceeds from the outer surface and time is required to conduct the heat away from the centre. The strain in the transformed martensite is tensile, in the circumferential direction, and so radial cracks form in the martensite to relieve the stress; these are known as *quenching cracks*. The properties of materials are therefore modified by the production of these residual stresses, whether of microscopic or macroscopic origin, and the behaviour of a structure cannot be interpreted correctly without taking the possibility of their existence into account.

10.4. *Corrosion*

The attack of a solid material by a solid, liquid, or gas, either removing portions of the original solid, or modifying its surface, can have many consequences. The solid–solid reaction between metals was discussed in Chapter 8 with reference to compatibility and diffusion. Here the concern is with solid–liquid and solid–gas reactions.

10.4.1. *The corrosion of metals*

Any elemental solid substance immersed in a liquid has a tendency for some of its atoms to enter the liquid in a mobile form. In most cases, the tendency is very small indeed, for only a few atoms could lower their free energy by so doing and these might be loosely bound ones at extremities on the surface. Any such atoms that did become free in the liquid would have a similar tendency to return to the solid, and so equilibrium would result, with a negligible solution of atoms of the solid present in the liquid. The measure of change in free energy in this process can be described as the '*solution potential*'. In the case of metals, another process can be added, that of oxidation, which does not imply merely the addition of oxygen, but the removal of electrons from an atom, leaving that atom as an *ion*. In the case of the best-known metal from the corrosion standpoint, iron, two types of ion are possible thus:

$$Fe \rightleftharpoons Fe^{2+} + 2e^- \quad \text{(ferrous ion } Fe^{2+}) \tag{10.1}$$

$$Fe \rightleftharpoons Fe^{3+} + 3e^- \quad \text{(ferric ion } Fe^{3+}) \tag{10.2}$$

The readiness to ionize by losing electrons in the presence of an ionizing solvent is called the *oxidation potential*. The reactions given are reversible, in that electrons can be picked up again to convert the ions back to atoms, and equilibrium between the forward and reverse reactions will soon be established. This equilibrium can be disturbed, either by the electrons being used up to add on to some other ions, or by combination of a negatively charged ion with the positively charged metal one. Both processes lead to an increase in the forward reaction, i.e. more iron goes into solution. If the electrons in equations (10.1) or (10.2) above are removed by the passage of an electric current, for instance, then iron will dissolve in an attempt to restore the balance, and so a continuous dissolution will result as long as the current flows. This feature is utilized in the electropolishing of metals, whereby a current is passed between the specimen and an inert electrode (often platinum) and the surface of the specimen dissolves.

If the metal ions meet a negative ion in the solution, such as the hydroxyl OH^- ion present in water, then they may combine, neutralizing the electric charges, and precipitating the metal hydroxide compound if this is insoluble in water. This is what happens in the rusting of iron, but here the oxygen in the air (or dissolved in the water) joins in, to create more hydroxyl ions by combining with the spare hydrogen in water, thus:

$$6H_2O \rightarrow 6H^+ + 6(OH)^-$$
$$3O_2 + 6H^+ + 6(OH)^- \rightarrow 12(OH^-) - 12e^-$$
$$4Fe \rightarrow 4Fe^{3+} + 12e^-$$
$$4Fe^{3+} + 12(OH^-) \rightarrow 4Fe(OH)_3$$

We may write, more simply, since all electric charges neutralize each other,

$$4Fe + 3O_2 + 6H_2O \rightarrow 4Fe(OH)_3 \qquad (10.3)$$

The ferric hydroxide produced in equation (10.3) is the familiar red-brown compound, iron rust.

The tendency for a metal to enter a solution (as ions, with the production of electrons) can be measured as an *electrode potential*, by comparing it to a standard hydrogen electrode, i.e. a supply of hydrogen tending to enter into a solution containing its own ions, $H_2 \rightleftharpoons 2H^+ + 2e^-$. (Of course, hydrogen cannot be handled like a solid; it is adsorbed on to platinum to provide a source of the atoms of the gas.) If a measuring instrument shows a positive *potential difference* when a metal is immersed in a solution containing a standard source of hydrogen, then the electrode potential of that metal is positive, because the metal is tending to enter the solution as ions more readily than is hydrogen. The metal is called the *anode*, and the hydrogen electrode in this case is called the *cathode*. All metals showing loose binding of their outer electrons will tend to dissolve (by ionizing) readily in a solution, since they easily shed those electrons. These metals therefore have a high positive electrode potential compared with hydrogen, e.g. lithium, with +2·96 V. All the alkali and alkaline earth metals show high electrode potentials (all contain *s* electrons). Hydrogen, on this scale, must be zero, since it is the datum line against which other potentials are measured. Some metals show a negative electrode potential, i.e. hydrogen enters the solution more readily than they do; these metals are *cathodic* to hydrogen, because they produce fewer electrons. Copper and the noble metals gold, silver, and platinum are cathodic to hydrogen (this is why they were called *noble* metals).

When two dissimilar metals are immersed in a solution, each will set up its own equilibrium between ions and electrons. The one with the higher electrode potential (both compared to hydrogen) will produce more electrons and more ions, and will therefore be anodic to the other. If they are connected externally by a wire, current will flow from anode to cathode in the external circuit. The cathode will now have an excess of electrons, upsetting its equilibrium. If the two metals are copper and zinc, immersed in sulphuric acid, electrode potentials are set up by each metal, thus,

$$Cu \rightleftharpoons Cu^{2+} + 2e^- \quad \text{Electrode potential about } -0{\cdot}34\ V$$
$$Zn \rightleftharpoons Zn^{2+} + 2e^- \quad \text{Electrode potential about } +0{\cdot}76\ V$$

(At the same time, water and acid both ionize to produce H^+ ions.)

The zinc has the higher electrode potential, so that electrons flow from the zinc (anode) through the external circuit to the copper. The equilibrium at the copper cathode is now upset, because more electrons are present than those normally produced by the solution of copper ions (see above). Some copper ions will therefore return to the metal, neutralizing some of the electrons by becoming atoms. The rate of the reverse reaction may alter so markedly that eventually insufficient copper ions become available to neutralize the electrons coming from the zinc externally. To neutralize these, hydrogen ions present in the water will be available, and hydrogen gas will then be produced at the cathode (copper). Meanwhile, the flow of electrons from the zinc through the external circuit has also upset the anodic equilibrium, and more zinc goes into solution to make up for it.

The net result of this complex process is the dissolution of the zinc anode, evolution of hydrogen at the cathode (copper), and a current flowing round the external circuit. This type of electrochemical cell, first demonstrated by GALVANI, leads to corrosion known either as *electrochemical* or *galvanic* corrosion. Placing dissimilar metals in external contact, both being immersed in an electrolyte, can lead only to dissolution of the one with the higher electrode potential, since it becomes the anode. The greater the difference in electrode potential between the two dissimilar metals, the greater the current flow (electrons) and therefore the more rapid the anodic dissolution. The greater the concentration of hydrogen ions in the electrolyte (more acidic) the more rapid is the neutralization of electrons by hydrogen ions at the cathode and the greater the evolution of hydrogen gas. In turn, the external current flow is more rapid once the cathode electrons are neutralized, and hence the corrosion of the anode increases by a feed-back mechanism. The higher the temperature, the more rapidly all reactions occur, and again the corrosion rate is increased.

A number of other factors can influence electrochemical corrosion. Inhibition or boosting of any of the reactions at the electrodes can affect the rate of attack on the anode. The increase in hydrogen-ion concentration referred to is one such effect. The transport of ions through the solution takes time, and increasing the distance between the electrodes will slow down the attack to some slight extent, whereas decreasing the distance may speed it up. If the flow of current through the external circuit is opposed, by a superimposed voltage from another source, then the whole effect may be reversed in such a cell. The supply of electrons to any electrode will tend to create a reversal of the dissolution process and cause a *plating* of the metal on to the electrode rather than the reverse.

It is not essential to have two dissimilar metals to form an electrochemical cell, although the wide possible difference in electrode potential with dissimilar metals obviously makes for rapid corrosion. If, in any one metal surface immersed in an electrolyte, there are areas of impurity, then a galvanic cell may be created by virtue of the varying electrode potential there. If the area is anodic to the rest of the surface, it will dissolve; if it is cathodic, the remaining surface will dissolve. Similarly, areas of lattice strain in the metal surface, even if absolutely homogeneous in composition, may become anodic to each other. Such areas are frequently produced by fabrication, machining, or polishing processes. The anodic areas again dissolve. In metallic alloys, therefore, where the composition of different phases may vary or the lattice strain, or both, a two-phase alloy will tend to corrode more readily because an electrochemical cell will be set up between the different areas. Mild steel, containing ferrite (pure iron) and pearlite (a eutectoid, itself two-phase), shows this, but in a very mild form, the ferrite being only slightly anodic to areas containing iron carbide (in pearlite). Some of the aluminium alloys, which can be hardened by precipitation of a compound as described in Chapter 8, are readily corroded in the two-phase (hardened) state by this same action. It is unfortunate that the condition for best mechanical properties in these and similar alloys often leads to greater corrosion than for the quenched or annealed states.

Fortunately, the mechanisms of electrochemical corrosion can sometimes be turned to advantage for protection of the component. The galvanizing of iron or steel, i.e. coating it with a layer of zinc, utilizes the higher electrode potential of zinc, compared to iron, to advantage. The zinc is anodic to the iron. If a break occurs in the coating, exposing the iron, and therefore the iron–zinc junction, the zinc dissolves in the electrolyte (rainwater mostly), and goes on dissolving until it is used up. Only then can the iron rust to any significant degree. The zinc here is acting as a *sacrificial anode* once the iron is exposed. Tinplate, on the other hand, is of value only so long as the coating remains intact, because tin is cathodic to iron. A scratch exposing the iron leads to iron becoming the anode and dissolving at a rapid rate, leaving the cathodic tin unaffected. The sacrificial anode can be used in many ways; it can be immersed in closed or open tanks, or fastened to the side of a structure. So long as there is electrical connection and an electrolyte, the anode must dissolve first. Magnesium is a reasonably cheap anode, used in water tanks and in the soil connected to buried pipes. Zinc has been used on ships' hulls and metal piers. The latest development uses platinized titanium; hydrogen is evolved and the anode does not dissolve. The use of an applied voltage to inhibit ionic transport or reactions at the electrodes has been referred to previously. Painting and coating, to keep the electrolyte away from the surface, are equally obvious remedies, though relying entirely on the physical separation of electrolyte from the electrodes.

When the dissolution of the anodic material is uniform, as occurs in many applications of materials such as chemical vessels for storage or manufacturing processes, few problems arise. There is however a danger when the electrochemical cell is set up by localized composition or stress changes, or the inhibition or boosting of some electrode reactions in local areas. This leads to so-called *pitting corrosion* when the anodic area is small compared with that of the cathode. This type of attack may occur in two-phase alloys when the dispersion of the anodic phase is very fine, or as a result of heat treatment or stress-raising conditions, as in welding. Local inhibition of reaction arises where oxygen plays an important part, as in the rusting of iron. Scale or dirt on an otherwise uniform iron surface may cause the production of anodic areas, the rust forming where oxygen has access, these sites becoming cathodic. If the dirt were removed, uniform oxygen access might no longer create such localized cathodic areas. Crevices and cracks also provide oxygen-inhibiting regions, and lead to exaggerated pitting corrosion, especially in the case of cracks. Grain boundaries in metals are less tightly bound areas of the lattice than the centre of grains, and therefore present a more ready source of ions and electrons, i.e. they may become anodic. Variations in orientation between one grain and another produce a similar result, a feature used in metallography when etching the grains of a metal specimen to show contrast in reflectivity under the microscope. Micro-organisms can affect corrosion, either by their influence on the chemical nature of the electrolyte, or by their own decomposition. Many of them produce, or consume, oxygen, and so can upset the equilibrium at the electrodes as mentioned in the case of dirt and scale.

10.4.2. *Oxidation in the absence of solvents*

All metals bear some oxide film on their surfaces, since the equilibrium pressure of oxygen above the surface (even for gold), at room temperature, is below the pressure of oxygen in the atmosphere. Oxygen must therefore add on to the surface of all metals in air or oxidizing atmospheres, either loosely, and involving only a few atoms (gold, platinum), or to form quite a thick, hard film (aluminium). The higher the temperature, the more rapid the reaction and the thicker the oxide film. The oxide film, however, acts as a barrier to further oxidation, since oxygen atoms in the atmosphere can no longer reach an uninterrupted metal surface. Either oxygen atoms must diffuse through the oxide film to meet the metal at the oxide/metal interface, or metal atoms must diffuse out through the oxide to meet oxygen at the oxide/atmosphere interface. (It is not impossible, either, that almost equal rates of diffusion of oxgyen and metal could occur, so that they would meet in the centre of the oxide film.) The metal ion being smaller than the oxygen ion can usually diffuse more readily through spaces in the oxide lattice (Figure 3.28 and Table IX). As the oxide film grows in thickness, the time taken for metal to diffuse through it, to continue the growth, must inevitably increase, and so cause the rate of growth to decrease. When this occurs, the growth rate depends on the square root of time, and hence a plot of oxide film thickness against time follows a parabolic form.

Occasionally, the metal ion occupies a much smaller volume than its atom, and hence the formation of oxide takes place with a decrease in net volume. This causes porosity in the oxide, and oxygen can readily penetrate the film and oxidize more metal on the surface. This was first demonstrated by Pilling and Bedworth, and accounts for oxidation occurring at a greater rate than the parabolic type that is more common. When the specific volume of oxide is less than that of the metal from which it forms (Pilling–Bedworth ratio < 1) the rate of oxidation may be linear, or it may increase in rate under certain circumstances. An increase in the rate requires a rise in temperature at the metal oxide interface (speeding up the reaction) owing to the free energy released, or a penetration of grain boundaries or other heterogeneities in the structure, eventually causing whole grains to become detached or surrounded by oxide, or dissolution of oxide in the metal. This type of oxidation may be termed *catastrophic*. Metals that show a decrease in volume in the formation of oxide are magnesium and most of the alkali metals; sodium, potassium, etc. Copper is a good example of a metal following a parabolic oxidation rate. Catastrophic oxidation occurs with molybdenum at high temperatures (1,000°C), owing to volatilization of the oxide, with niobium and tantalum owing to solution of oxygen in the metal, and with uranium owing to a summation of similar causes.

The structure of the oxide in relation to the metal substrate is obviously important. If the lattice of the oxide can match up with that of the metal on which it forms, i.e. if some atoms are shared by each, then the coherent structure that results is quite firmly bound to the substrate and will not readily be detached. The void space in the oxide lattice is then also likely to be less, and so the rate of diffusion will be less, particularly if the bonding between metal and oxygen in the oxide is also very strong, as with aluminium and chromium. Where the oxide is non-coherent, especially if the metal–oxygen binding is also low, then rapid oxidation will occur. It is obvious that where the specific volume of the oxide is less than that of the metal then coherency is almost impossible, as in magnesium, where the oxide occupies only 80% of the volume of the metal from which it is formed.

High temperatures will speed up the rate of oxidation reactions, since diffusion rates are increased, and the extra energy causes more rapid chemical combination of metal and oxygen. Oxygen diffusion increases in addition to metal diffusion, and so where at lower temperatures a parabolic rate, dependent on metal diffusion as the controlling factor, occurs, changes may take place at higher temperature. First the slope of the parabola steepens, then there is a gradual transition to a linear rate as oxygen penetration takes the dominant role. Finally, oxygen penetration of grain boundaries or around inclusions, or a second phase, may become predominant, and catastrophic oxidation sets in. These changes are aggravated by secondary reactions, such as the decarburization of steel, whereby iron carbide Fe_3C is attacked by oxygen more readily than the ferrite matrix. Occasionally, the secondary reactions or selective oxidation may be of value in improving oxidation resistance by providing a protective coating. The classic case is that of austenitic steels containing a high percentage of chromium. Although the oxide film contains all the major elements (Fe, Cr, Ni) present, it is the oxide of chromium Cr_2O_3 that most resists further diffusion of metal or oxygen, and probably forms something approaching a spinel-type structure with the iron and nickel oxides (spinel oxides are discussed in Chapter 3, page 73). At very high temperatures the structure of this oxide is no longer so protective, and oxygen diffusion commences, to attack grain boundaries and produce rapid scaling.

Mention has already been made of molybdenum undergoing catastrophic oxidation. One of the main drawbacks to the use of the transition metals of high melting point (which retain high strength at high temperatures) in industry is still the problem of oxidation. Molybdenum can be coated with its compound with silicon $MoSi_2$ and the molybdenum volatilizes away as MoO_3. The silicon, however, forms a coherent layer of silica SiO_2 which continues to give protection so long as it remains coherent. Even a humble material such as cast iron has been used for at least three centuries in firegrates because of the protection afforded by a film of silica that forms on the surface. Silicon steels, used for internal-combustion engine valves, were developed primarily for their oxidation resistance, particularly when chromium was also incorporated to provide strength and a further contribution to a resistant oxide film. Attempts to produce glass-type coatings are only partially successful, the difficulty being that of thermal shock, even where coherent binding can be obtained. Intermetallic compounds produced by diffusion may be the next development.

Occasionally the rapid oxidation of metals can be utilized to advantage, as in vacuum systems, to *getter* the oxygen remaining after prolonged pumping. Electronic valves and tubes contain a small capsule sealed into the tube, which is fired electrically after pumping and sealing, to remove the last traces of oxygen. In these applications, the oxide must not flake away from the metal surface, and must not volatilize, since contamination of components may be detrimental. Zirconium is an excellent getter for oxygen, for the oxygen tends to dissolve in the metal and any film remains tightly bound to the surface.

The passive nature of the surface of a metal on which a stable coherent oxide film has been built up can also be used to establish resistance to liquids, where electrochemical corrosion might otherwise result. The stable coherent oxide films have a low oxidation potential, since they are electrically neutral, and so both metal and oxygen ions tend to enter the solution together. To free electrons, as with a metal alone, a change in the electronic system of the oxide must take place, or metal must diffuse through the oxide to the liquid, or liquid through the oxide to the metal. Of these possibilities, the second is most likely, and this may be very slow, as in the case of aluminium, chromium, and silicon oxide films on metals or alloys. Where the liquid is reducing rather than oxidizing, then the pattern may change, and an oxide film may no longer prove protective. Aluminium is frequently *anodized*, i.e. treated in an electrochemical cell to build up a thick coherent layer of oxide, which makes it extremely passive to oxidizing acids and atmospheres. The oxide film on chromium-bearing austenitic steels is resistant to oxidizing liquids as well as to high-temperature oxidation. Nitride and phosphate films on iron convey some degree of passivity by providing a barrier to metal diffusion.

Attack of metal surfaces at high temperatures can take place in atmospheres other than oxygen. One of the most deleterious is that of sulphurous gases. These not only oxidize the surface, but in many cases also form sulphides of the metals. The metals and alloys that normally form stable, coherent oxide films in oxygen-bearing atmospheres also do so in the presence of sulphur dioxide and trioxide, but usually at a rather more rapid rate of reaction, and with a lower temperature for change-over from parabolic to linear rate. Sulphur, present at the metal/film interface, can preferentially attack grain boundaries, in some cases more potently than oxygen; nickel and its alloys are so attacked. The grain boundaries tend to form brittle sulphide films in such cases, with considerable reduction of mechanical strength. More mildly oxidizing gases, such as carbon dioxide, can under certain circumstances form carbonaceous deposits, which on steels may carburize the material and render it more brittle, but in a more uniform manner than in the case of sulphur. Hydrogen sulphide, a reducing gas, will reduce the oxide film and cause rapid attack by sulphur and is therefore to be avoided. All these gases tend to be present, to varying degrees, in flue gases at high temperatures.

10.4.3. *Dissolution by solvents*

Apart from the electrochemical corrosion of metals, the tendency for any substance to lose atoms or molecules into a liquid medium in which it is immersed is important for non-metallic materials. Normally, the criterion for dissolution is the relative sizes of the solid ions or molecules compared with those of the liquid. Compounds also tend to dissolve in solvents having a similar type of structure (as with metals entering into solid solution). Thus, paraffin wax will readily dissolve in liquid hydrocarbons of paraffin type but not of the aromatic variety. Oxides of metals dissolve in melts containing other oxides or silicates, but not so readily, or not at all, in liquid metals.

So far as the size effect is concerned, the complexity of some molecular groupings affects the relative size factor. Sulphur ions are larger than oxygen ions, and sulphides of metals are not normally so readily soluble in acids as the oxides of the metals. Silicates are complex but stable compounds, and do not dissolve so readily in acids as oxides, halides, or sulphides. If the complex molecule can be broken down, as in the degradation of polymers, then dissolution in a solvent becomes more rapid.

Rise in temperature affects these processes by the speeding-up of atomic or molecular vibrations and by causing more rapid dissolution, i.e. separation of small units of solid to enter void spaces in the structure of the solvent. The general rule is that corrosion increases by a factor of 2 for every 10 degC rise in temperature. This rule applies only to general dissolution, and does not account for any possible change in the mechanism of attack, so that it must be used with discretion.

Oxides and similar compounds of metals may ionize partly when entering a solution, and some electrical conductivity may result, but this aspect of solution is really outside the scope of this book. The selection of a solvent for a solid substance can thus be made on the basis of similar structure and similar size in the first instance. (The principles were explained for metals in more detail in Chapter 8.)

10.4.4. *Corrosion and mechanical properties*

The environment of a material may markedly influence its behaviour in an engineering capacity. Unless this is taken into account, the choice of materials purely for mechanical strength or other physical properties may result in premature failure or inadequate performance. The presence of a corrosive environment can emphasize mechanical failure by two main processes, those of *stress-corrosion*, and *corrosion fatigue*. Other phenomena have been described, but are possibly special cases of these two. In stress-corrosion, the liquid or gaseous medium may accelerate the propagation of cracks formed by stress, or the existence of stress may accelerate corrosion at localized areas. In either case, the result is an acceleration of failure. Corrosion fatigue, on the other hand, is simply an acceleration of the fatigue process produced by the environment, leading to more rapid failure under the same stress conditions, or failure at a lower alternating stress in the same time. The environments producing either of these types of failure are fairly specific to particular materials. High polymers are affected by acidic oxides (SO_2, CO_2), sea-water, and conditions of radiation and electrical stress. Mild steel is affected by caustic alkalis, as in boiler water; brasses by chlorides; *stainless* steel by chlorides; nickel alloys by sulphurous gases; to name but a few.

Fatigue and other stress conditions can be affected by irregularities in the surface such as machining marks, exaggerated changes in cross-section, stamp marks, and any similar stress concentrations. A corrosive environment must further exaggerate these effects, and increase the liability to failure. True corrosion fatigue reduces the fatigue life even when such surface defects are absent, and should be distinguished from the more obvious influence of stress concentrations.

10.5. *Radiation damage*

The advent of the nuclear energy industry has raised another important service factor in the behaviour of materials. The majority of radiations encountered previously (X-rays; electron beams (β-rays); radio waves) produced no permanent effects on materials in their path, other than a rise in temperature. The short range of α-particles (helium nuclei) and β-rays implied a very small penetration into all substances. The X-rays, being of equivalent shorter wavelength, could penetrate farther, and γ-rays of even shorter wavelength, farther still. Nevertheless, only biological tissue appeared to be unduly affected by these radiations, and engineering materials were, and still are, used to shield living tissue from them, by absorbing the radiation. Heavy metals absorb these better than light metals, and all metals are better than comparatively open structures such as wood and brick. Barium absorbs better than calcium, and so barytes–concrete blocks were produced, containing a high proportion of barium, for building purposes.

The nuclear energy developments emphasized a far more penetrating particle, the neutron, which, being electrically neutral, can move freely through an atomic lattice of any material. The life-span of the neutron depends on its kinetic energy, and on the material through which it passes. Collisions occur with the atomic nuclei and a change in momentum is involved. Some measure of the life-span of the neutron is required, and the term *nuclear cross-section* is used to define the effective area that the target nucleus presents to the neutron as it moves through the lattice, on a statistical or probability basis. An atomic diameter of about 2 ångström (2×10^{-8} cm) means that the cross-sectional area of the complete atom is of the order of 10^{-15} cm^2, of which the nucleus occupies only a fraction. The unit of nuclear cross-section used is called the *barn* (bn), equivalent to

10^{-24} cm^2, so that a cross-section of a few barns implies a nucleus of such small dimensions that the neutron will travel through 10^8 atoms on average before it is likely to strike a nucleus. The mean free path of neutrons in most solids will therefore be a few centimetres in length.

The neutron must pass through the electron cloud of some of the atoms, and may interact with electrons. These may be displaced in energy level, or even knocked to infinity, producing an ion. Secondary effects may result from the displaced electrons, and certainly heat will be developed. When the neutron subsequently collides with a nucleus, as it must, it may be deflected by an equivalent mechanical collision (like two billiard balls) or absorbed into the nucleus, which already contains neutrons (like two waterdrops coalescing). In certain special cases, the absorption may overload the stability of the nucleus, and nuclear fission results. In the first case, collision with change in momentum may displace the whole atom, or merely result in slight deflection. This probability is referred to as the *scattering cross-section*. In the second case, provided that the atom is not unstable as a result, the condition of probability is the *capture* or *absorption cross-section*. In the third case, the *fission cross-section* is concerned. Capture, with or without fission, is affected not only by the probability of a collision, but also by particular resonance effects, such that neutrons of particular kinetic energy values are more likely to be captured. The probability of neutron scatter or capture is not only dependent on the size of the target, therefore, but also on the speed with which the neutron moves. As a result of scattering collisions, the neutron energy decreases towards the end of its free path, and so capture by resonance at lower energy may become more probable. This factor is utilized in nuclear reactors, where a *moderator*, or slowing-down material, is used to decrease the energy of fast neutrons produced in fission until they are more readily captured in the fissile material at lower energies. This is of particular value when using natural uranium, which tends to capture fast neutrons in the ^{238}U isotope rather than cause fission in the very small number (0·7%) of fissile ^{235}U atoms. (Fast neutrons may be moving at speeds of the order of 10^{10} cm sec^{-1} with an energy of 1 to 10 MeV. When slowed down to a similar order of energy to that of the thermal vibration of atoms (hence called *thermal neutrons*) the speed is about 10^5 cm sec^{-1} at an energy of 1 eV (1 eV = $1{\cdot}6 \times 10^{-12}$ ergs).) The nuclear cross-section varies, therefore, with neutron energy. Capture inevitably gives rise to some secondary effects, either release of heat energy or the emission of electrons (β- or γ-rays). Any absorber or moderator will therefore suffer rise in temperature, and perhaps displacement of atoms, as a result of neutron bombardment. The energy transmitted by a neutron collision with an atom may be 10^5 eV, and on the average one collision takes place in every 500 atoms through which the neutron moves. The energy required to displace an atom is about 25 eV, so that even one neutron–atom collision can send the atom careering through the lattice, itself colliding with other atoms, disrupting the lattice, and releasing heat energy in the process. The path of such an atom is referred to as a *radiation* or *displacement spike*, and the damage caused as *knock-on damage* (Figure 10.1).

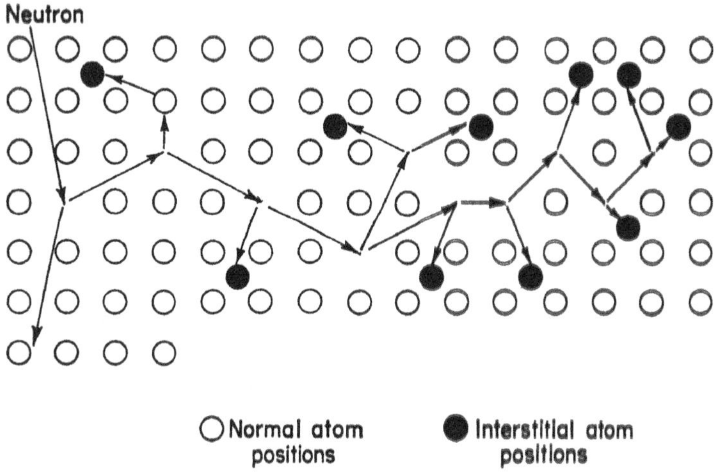

FIG. 10.1. Two-dimensional representation of a radiation spike.

Excitation of electrons by the passage of neutrons has little effect on materials in which the electron cloud behaves more like a fluid, as in the metals. The good conductors of heat and electricity all rely on electron movements in some form or another, and so are little affected by such electron displacements. In other materials, where covalent or homopolar bonds are involved, the electrons in the higher energy levels are shared in some way to produce the binding of the atoms. The bond therefore may be affected by excitation of these electrons. In inorganic salts, with ionic bonds, defects in the lattice may be produced as a result. In organic (covalent) compounds, the covalent bond may be shattered, thus *cracking* the molecule into two smaller units. The long-chain polymers may be broken up into smaller molecules, with obviously different properties. The material *Perspex*† can be so degraded, bubbles of gas appearing

† Registered trade mark.

inside the transparent material when bombarded with neutrons. On the other hand, side-chain polymers may suffer increased activation, the side-chain groups linking together to form an even larger molecule. Cross-linking of this type can occur with polyethylene, and the resultant larger molecules are harder, more brittle, and of higher melting point. Lubricants based on paraffin-type hydrocarbons may be degraded into smaller molecules, but aromatic compounds are more resistant. Some aromatic ring compounds have been used for

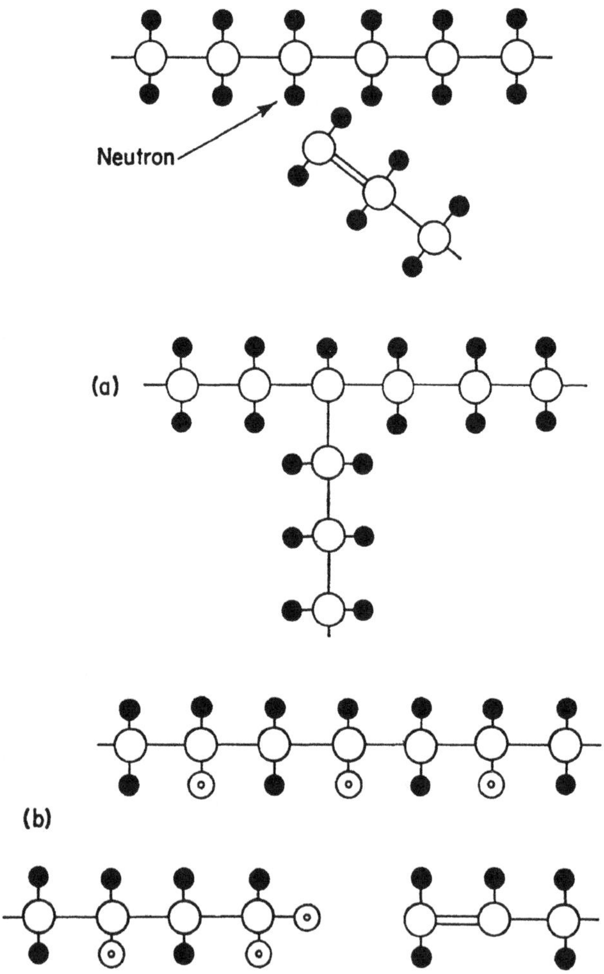

FIG. 10.2. (*a*) Cross-linking of polyethylene. (*b*) Degradation of polyvinyl chloride.

this reason as neutron moderators in nuclear reactors, e.g. the phenyl compounds. Even these break down eventually, and have to be replaced. Figure 10.2 illustrates both cross-linking and degradation of polymers.

Knock-on damage in a radiation spike may push atoms into interstitial positions that normally they cannot occupy, and hence give rise to strain in the lattice. Vacant lattice sites will be created by the removal of these atoms from their normal positions. A number of defects may arise from the impact of one neutron, since the knock-on atom may move some distance, as shown in Figure 10.1. The probability of more atoms being involved increases with neutron energy, since the number of collisions goes up. The cluster of defects left by a radiation spike extends over a distance of about 10^{-4}–10^{-6} cm in each direction. The atoms are also affected by the thermal energy released, and so vibrate more than usual in the vicinity of the radiation spike. This gives rise to a greater probability of movement among the point defects created, to restore the original conditions and so relieve some of the lattice strain involved. Movement between an interstitial atom and a vacancy can lead to mutual annihilation, and a consequent restoration of the normal regular lattice. This annealing process therefore accompanies the radiation damage but inevitably lags behind it, since diffusion processes are much slower than the neutron passage, and the energies involved are vastly different. In many cases, however, the damage to a metal may have completely annealed out by the time the specimen has been removed from the neutron source and tested. Aluminium (f.c.c.) anneals rapidly provided that the temperature exceeds −50°C, whereas iron (b.c.c.) requires an annealing temperature greater than 300°C.

In lattices of greater complexity than most metals (oxides, silicates) the damage is not annealed so readily. Anisotropic materials, e.g. the hexagonal metals and graphite, may suffer damage resulting in changes that are non-uniform in direction, and so irreversible. This may lead to permanent changes in shape. Fission results in two atoms replacing the site of an original one, and the release of tremendous energy, of the order of 200 MeV. The *fission products* may be liquid or gaseous, and so modify the properties of the fissile material once enough atoms have been fissioned. The gases, in particular, are frequently those of the rare gas group, particularly xenon, and can cause swelling of the material if they coalesce to form bubbles, since the temperature is high and the material reasonably plastic. If the basic material is non-plastic, the bubbles may produce severe cracking.

Complex lattices such as those of the silicates suffer damage when present in the form of concrete biological shielding around a nuclear reactor. Quartz is converted to an amorphous lattice by the displacement of the ions by neutrons, and silicates suffer a similar fate in time. The cement bond in concrete therefore tends to disintegrate unless the neutron flux is kept low. Other heterogeneous materials such as wood, brick, and stone react in a similar manner.

Metals, as already explained, may anneal the damage. The hardening produced is not, however, always completely removed. In the case of body-centred iron (mild or low-alloy steel) the transition

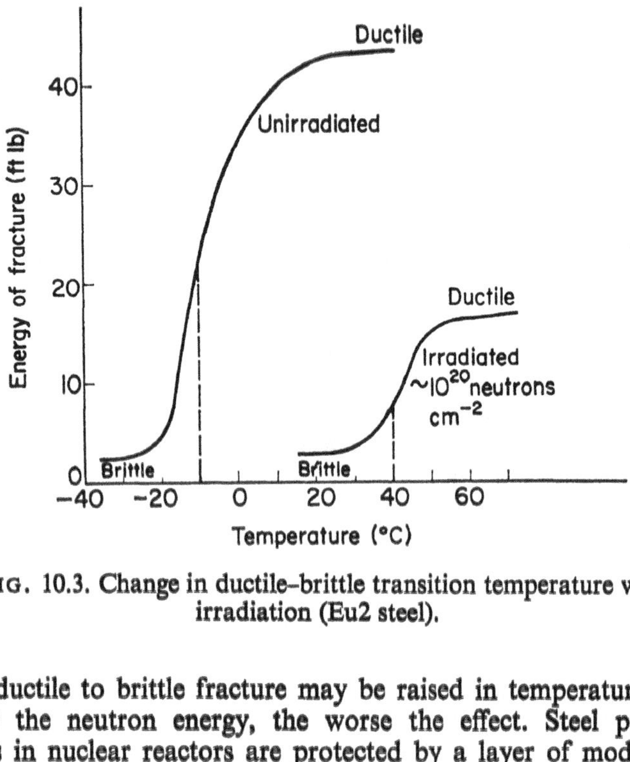

FIG. 10.3. Change in ductile–brittle transition temperature with irradiation (Eu2 steel).

from ductile to brittle fracture may be raised in temperature. The higher the neutron energy, the worse the effect. Steel pressure vessels in nuclear reactors are protected by a layer of moderator, so that more of the thermal energy neutrons, and less of the faster ones, reach the steel. Nevertheless, considerable concern arises from this problem. If the metal is raised in temperature, in order to anneal out the damage, then it cannot support the operational stresses so well, and may creep, with dangerous implications of subsequent failure. If a more highly alloyed steel is used, then the creep strength goes up, but unfortunately so does the possibility of a higher brittle transition temperature. The main problem arises from the decreased ductility that may accompany the rise in yield strength, and yield strength may increase by 10 ton in^{-2} in a thermal reactor operated for 10 years at an average fast neutron flux of 10^{12} neutrons cm^{-2} sec^{-1}. In mild steel, each rise in yield stress of 1 ton in^{-2} produces 3 degC or so rise in brittle transition temperature (Figure 10.3).

Graphite has been extensively used as a moderator in nuclear reactors. The hexagonal lattice of graphite being anisotropic, the knock-on damage is not easily annealed out, and a moderator receives the high-energy neutrons immediately after fission. The strain in the graphite lattice irradiated by neutrons is very high, and considerable energy is stored in the lattice as a result. This energy can be released only by increase in thermal energy, to activate the atoms and enable the interstitial atoms to return to normal sites. During reactor operation the temperature is high enough, but the continuous neutron flux causes the rate of damage to exceed by far the diffusion process of annealing. Removal of stored energy must therefore be carried out in the absence of the neutron flux. This stored energy can be so enormous that the release per unit mass of material, expressed as heat energy, may exceed the specific heat of the material, leading to a further increase in temperature, and thus to a more rapid release of stored energy and to a continued temperature rise. The process must obviously be very carefully controlled. The stored energy is named, after the original discoverer, *Wigner energy*, and the removal referred to as *Wigner release*. Catastrophic temperature-rise can occur in Wigner release, as exemplified by the Windscale reactor accident in 1957 (Figure 10.4).

Chemical reactions are affected by neutron irradiation, as might be expected from the effects on solid materials. Apart from speeding up reactions by local rise in temperature, neutrons can activate the reactants as described for organic and inorganic compounds. The nature of reactions may change, but the kinetics are often altered out of all proportion. This is usually due to the creation of *active species*, i.e. ionic or molecular groupings that cannot exist under normal conditions but in the circumstances of high-energy bombardment may exist long enough to change the type or progress of a reaction. Liquids such as water (light or heavy) and gases such as carbon dioxide may be used as coolants in nuclear reactors, and so subject to irradiation by neutrons. In the presence of graphite, carbon dioxide containing a little moisture is believed to form metastable hydrocarbons, which cause a sequence of different chemical reactions to occur with graphite, removing more than normal amounts into the gas phase, and depositing it elsewhere, in cooler regions or away from the neutron source. Metals forming normally stable oxide films, passive to water, may become corroded by a similar chain of events.

Other phenomena are affected by neutrons, notably the mechanical strain produced by dislocation movements. Creep takes place at an increasing rate under irradiation, or at a lower stress than would

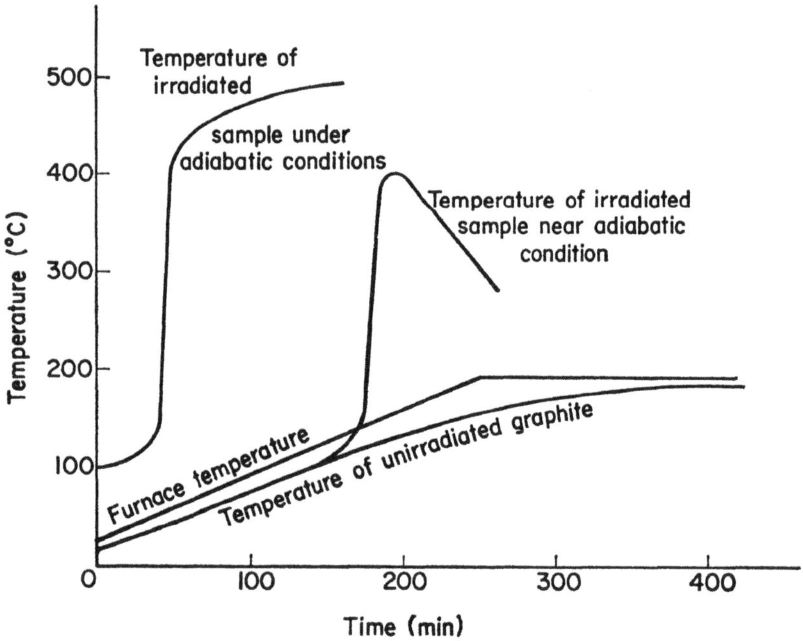

FIG. 10.4. Release of stored energy in graphite.

normally be expected. Neutron irradiation can thus seriously affect the properties and behaviour of many engineering materials, and this accounts for some of the complexity of nuclear installations, particularly of those components operating within the influence of the neutron flux.

10.6. *Conclusion*

The summation of the service factors applicable to particular environments or components is in the realm of technology and design rather than science. Empirical experiments may sometimes be necessary to evaluate performance data but, in general, more valuable information must arise from a scientific appraisal of what really constitutes the problem and an analysis of the individual mechanisms responsible. In the main, empirical experiments give one answer to one problem, whereas a fundamental understanding frequently enables some future set of conditions to be considered without further tests other than an extension to appropriate ranges of temperature, time, stress, heat flux or magnetic field, etc.

Further Reading

Materials science

VAN VLACK, L. H., *Elements of Materials Science,* Addison-Wesley, Reading, Mass., 1959.

EVANS, E. C., *An Introduction to Crystal Chemistry,* Cambridge University Press, London, 1948.

SCHMIDT, A. X., and MARLIES, C. A., *Principles of High Polymer Theory and Practice,* McGraw-Hill, New York, 1948.

Solid-state physics

ZWIKKER, C., *Physical Properties of Solid Materials,* Pergamon Press, London; Oxford, 1955.

HUTCHINSON, T. S., and BAIRD, D. C., *The Physics of Engineering Solids,* John Wiley, New York, 1963.

SEITZ, F., *Physics of Metals,* McGraw-Hill, New York, 1943.

Metallurgy

HUME-ROTHERY, W., *Elements of Structural Metallurgy,* Institute of Metals, London, 1961.

GUY, A. G., *Elements of Physical Metallurgy,* Addison-Wesley, Reading, Mass., 1959.

Non-metallic materials

NORTON, F. H., *Elements of Ceramics,* Addison-Wesley, Reading, Mass., 1952.

STANWORTH, J. E., *Physical Properties of Glass,* Clarendon Press, Oxford, 1950.

SIMONDS, H. R., WEITH, A. J., and BIGELOW, M. H., *Handbook of Plastics,* Van Nostrand, New York, 1949.

KEYSER, C. A., *Materials of Engineering,* Prentice-Hall, New York, 1956.

Author Index

Subject Index

www.ingramcontent.com/pod-product-compliance
Lightning Source LLC
LaVergne TN
LVHW020630300926
849161LV00004B/654